METHANE GENERATION FROM HUMAN, ANIMAL, AND AGRICULTURAL WASTES

Report of an Ad Hoc Panel of the Advisory Committee
   on Technology Innovation
Board on Science and Technology for International Development
Commission on International Relations
National Research Council

Books for Business
New York, Hong Kong

Methane Generation from Human, Animal, and Agricultural Wastes

by
National Academy of Sciences

ISBN 0-89499-020-9

Reprinted from the 1977 edition

Books for Business
New York, Hong Kong
http://www.BusinessBooksInternational.com

Panel on Methane Generation

GERARD A. ROHLICH, Department of Civil Engineering, Lyndon B. Johnson School, University of Texas, Austin, Texas 89812, USA, *Cochairperson*

VIRGINIA WALBOT, Department of Biology, Washington University, St. Louis, Missouri 63130, USA, *Cochairperson*

LARRY JEAN CONNOR, Department of Agricultural Economics, Michigan State University, East Lansing, Michigan 48823, USA

CLARENCE G. GOLUEKE, Department of Civil Engineering, Sanitary Engineering Research Laboratory, Richmond Field Station, University of California, Richmond, California 94804, USA

THOMAS D. HINESLY, Department of Agronomy, University of Illinois, Urbana, Illinois 61801, USA

P. H. JONES, Institute of Environmental Sciences and Engineering, University of Toronto, Toronto, Ontario, Canada

HERBERT M. LAPP, Department of Agricultural Engineering, University of Manitoba, Winnipeg, Manitoba, Canada

RAYMOND C. LOEHR, Department of Civil and Environmental Engineering, Cornell University, Ithaca, New York 14850, USA

CECIL LUE-HING, Metropolitan Sanitary District of Greater Chicago, 100 East Erie Street, Chicago, Illinois 60611, USA

JOHN T. PFEFFER, Department of Civil Engineering, University of Illinois, Urbana, Illinois 61801, USA

T. B. S. PRAKASAM, Metropolitan Sanitary District of Greater Chicago, Research and Development Laboratory, 5915 West Pershing Road, Chicago, Illinois 60650, USA

NORMAN L. BROWN, Board on Science and Technology for International Development, Commission on International Relations, National Academy of Sciences–National Research Council, 2101 Constitution Avenue, N.W., Washington, D.C. 20418, USA, *Staff Study Director*

JULIEN ENGEL, *Head, Special Studies*

Preface

In recent years there has been growing interest in the use, as a fuel, of methane generated by the decomposition of organic matter under conditions where contact with oxygen is eliminated. This interest has been sparked by a number of developments, not least of which is the rapid rise in price of fossil fuels and the particular burden this places on developing countries.

A wide variety of popular, semitechnical, and technical articles has appeared, based largely on experience in India. The articles are concerned with adoption of the technology by environmentally concerned individuals in industrialized nations as an alternative to conventional energy sources, particularly in rural situations. Except for a few articles—written primarily in India—that discuss rural use in particular countries, most of them seem to assume the existence of a technological infrastructure that can supply the equipment, hardware, and plumbing fittings that provide much of the convenience of the systems described. Furthermore, the articles often fail to deal adequately with the fact that these are biological systems, with the inherent chemical and biological problems frequently encountered in such systems.

The Advisory Committee on Technology Innovation (ACTI) through the Commission on International Relations and its Board on Science and Technology for International Development (BOSTID) believed that it would be helpful at this time to prepare a compendium of up-to-date information on the subject that would provide 1) general background for officials in developing countries who are responsible for rural planning and development, and 2) technical information for those interested in undertaking methane-generating projects in rural areas.

This study is written as a companion piece to an ACTI report on renewable resources of energy, and technologies for their exploitation, in rural areas of developing countries.* The generation of methane from human, animal, and agricultural wastes for use in these rural areas appeared to warrant special, intensive analysis in a publication of its own. The panel of experts

*Energy for Rural Development: Renewable Resources and Alternative Technologies for Developing Countries. Washington, D.C. 1976: National Academy of Sciences.

v

appointed for this purpose was asked to assess the current state of knowledge and technology, to suggest useful applications, and to identify areas where research and development are needed. The panel was requested to bear in mind that the report is directed to a dual audience: the technologist familiar with some of the basic principles involved—but not necessarily with inherent problems, recent developments, or information sources—and the decision maker and planner who must evaluate technical proposals in this field on the basis of a country's needs and constraints.

This report discusses the means by which the natural process of anaerobic fermentation can be controlled by man for his benefit, and how the methane generated by this process can be used as fuel. The report includes a bibliography of more detailed works on the subject and a list of the panel members to whom technologists in developing countries may turn for direct assistance with specific problems, if necessary.

It is the panel's hope that this report will not only meet a current need for information on the use of anaerobic fermentation of waste materials to produce fuel and fertilizer, but that it will also stimulate interest in using this technological approach to improve the quality of rural life.

Acknowledgments

The Panel on Methane Generation from Human, Animal, and Agricultural Wastes received help from many sources and wishes to thank all who have contributed to this report.

For their contribution to the initial discussions during the organization of this study, the panel expresses its appreciation to Professor Milton Barnett, Mr. Arjun Makhijani, and Dr. Thressa C. Stadtman.

For information on crop residues, thanks are due to Mr. Alan D. Poole, Institute for Energy Analysis, Washington, D.C. Dr. D. D. Schulte (Assistant Professor), and Mr. E. J. Kroeker (Research Associate), Department of Agricultural Engineering, University of Manitoba, Winnipeg, Manitoba, provided assistance in discussions of performance measurement. In addition, Dr. Schulte's help in discussions of operation, maintenance, and safety is appreciated.

Comments and criticisms of those who reviewed the manuscript have been helpful in eliminating ambiguities that crept into some of the discussions and some details that detracted from more essential material. For this, thanks are due Professors Robert M. Walker (Washington University, St. Louis, Mo.), Franklin R. Long (Cornell University, Ithaca, N.Y.), and Roger Revelle (University of California at San Diego, California).

To F. R. Ruskin, to whose lot fell the task of editing what must have seemed at times somewhat turgid prose dealing with materials not generally discussed in such detail in Academy reports, the Staff Study Director expresses his appreciation for the vast improvements in clarity and lucidity that resulted from her efforts.

Finally, on behalf of the Staff Study Director, special thanks are due Dr. T. B. S. Prakasam for his constant willingness to step into the breach and supply information that seemed unobtainable from any other source. Much of the practical engineering detail and many of the references to work in India would not have been included in this report without his cooperation.

Symbols

Btu	British thermal unit
cm	centimeter
cm^2	square centimeter
°C	degrees Celsius
°F	degrees Fahrenheit
ft	feet
ft^2	square feet
ft^3	cubic feet
hr	hour
J	joule
kg	kilogram
kJ	kilojoule
kWh	kilowatt hour
l	liter
lb	pound
m	meter
m^2	square meter
m^3	cubic meter
mg	milligram
min	minute
ml	milliliter
mm	millimeter
ppm	parts per million
psi	pounds per square inch
psia	pounds per square inch absolute

Contents

Tables

PART I

OVERVIEW

Chapter 1

Introduction and History

The need for alternative energy supplies is increasingly apparent. Our fossil fuel reserves will eventually be exhausted. Moreover, these reserves are unequally distributed and are becoming too costly for many countries that must purchase them. In addition, the cost of transportation may sharply limit the use of fossil fuels in the rural areas of many developing countries. And, as recent events have shown, the cost—and the availability—of these fuels is determined less by market forces than by the decisions of the producing nations.

The present energy-generating systems in developing countries depend largely on local resources: wood, straw, or dung for burning; hydraulic power for water wheels and electric power generation; and whatever fossil fuel supplies are locally available. A country's energy requirements often are not fully met by these local resources, and foreign-currency reserves must be expended to import the needed fossil fuel. In most developing countries, the economic base and the majority of the population are still rural, and machinery that requires energy (especially fossil fuel) is not heavily utilized. However, the lack of cheap and adequate energy often hampers rural development plans and retards improvement in the quality of rural life. Solving the problems of energy generation and distribution is central to implementation of plans for economic development, especially in rural areas. As solutions dependent on imported fossil fuels become increasingly expensive, the rewards of developing alternative fuel supplies from local materials are bound to grow.

This report is devoted to the development of an alternative energy resource suitable for individual or village use in a rural environment. An ideal resource is one that is local in origin and can produce energy useful for this purpose depending only on local materials and labor. Unlike present rural energy generation (primarily heat from burning), an ideal system should involve a fuel that can be utilized for different kinds of work. The ideal fuel resource should be easily accumulated and stored to provide energy for heating, lighting, small-scale electric power generation, and power for engines as needed. Moreover, where possible, it should also provide more energy than is now obtained from the same materials.

3

Rural areas usually have large supplies of material—crop residues and animal wastes—theoretically suitable for conversion into a usable source of energy. The process that appears to hold the greatest immediate potential for utilization of these materials as sources of fuel is anaerobic fermentation. This process, also called anaerobic digestion, converts complex organic matter to methane and other gases. It has advantages that recommend it for serious consideration: 1) it is the simplest and most practical method known for treating human and animal wastes to minimize the public health hazard associated with their handling and disposal; and 2) the residue left after removal of the gas is a valuable fertilizer that contains all the essential nutrients present in the raw materials.

The fact that organic material, rotting under conditions where it is out of contact with air, will produce a flammable gas has been known for centuries, particularly in the phenomenon of marsh gas. The occasional dancing flames of this gas (ignited, perhaps, by stray sparks from a nearby fire), seen at night, have given rise to the legends of the "will-o'-the-wisp," or fool's fire. Although it is not certain when it was first recognized that manure, if allowed to decompose in a sealed pit, would also produce a flammable gas, we know that the gas from a "carefully designed" septic tank was used for street lighting in Exeter, England, in 1895.[1] The experience must have been successful enough to encourage others, for, in the 1920s, several devices were built and used in England, specifically for the purpose of generating this gas,[2] which is primarily methane, the simplest organic compound of carbon and hydrogen. The process has also been utilized where energy supplies have been reduced, as in France, Algeria, and Germany during and after World War II, when methane thus produced was used to run automobiles.

In countries hampered by low natural abundance or inadequate distribution of energy supplies, methane-generating equipment has often been adapted to meet rural needs. Family-size methane-generating units have been used in diverse climates and cultures. In India, concern over the loss of cow dung for fertilizer, because of its traditional use as fuel, sparked early experiments to develop a system to provide fuel without destroying the dried dung. These experiments were initiated in 1939 at the Agricultural Research Institute in New Delhi.[3] An account of that experience states: "The experiments resulted in the designing of a simple and easy-to-operate gas plant in which dung is fermented to yield a combustible gas which can be used as a fuel and the dung residue . . . can be utilized as manure."[4] The work in India continued and expanded with the encouragement of the Khadi and Village Industries Commission. In 1961 the Gobar Gas Research Station was started in Ajitmal, Etawah (Uttar Pradesh), and in 1971 it published a variety of designs for gas plants. In the years since experiments first began in India, many thousands of such plants have been built in that country—most of them in rural areas and serving from one to several families.

The interest in this anaerobic digestion process is not as easily chronicled for other developing countries as it is for India, with the exception, perhaps, of Taiwan. There, experiments with the generation of working fuel from pig manure began about 1955 and developed into a program supported by the government.[5] To date, about 7,500 such devices have been built in Taiwan as permanent adjuncts to pig-raising operations on small- and medium-size farms, although reports indicate that perhaps only half that number are actually in operation.[6]

In the People's Republic of China (PRC), the practice has been promoted vigorously, at least since 1970. Several symposia on the subject have been held, and in 1972 a joint meeting of the Chinese Academy of Science and the Ministry of Agriculture and Forestry was held in Szechwan to summarize progress.[7] Available reports indicate that tens of thousands of methane generators are operating in the PRC, based on the use of night soil and other manures as raw materials. In Szechwan Province alone, for example, more than 30,000 small-scale generating installations have been reported.[8]

Korea, too, has wide experience with such rural units, with 24,000 having been installed between 1969 and 1973.[9]

There are scattered reports of the use of methane generation from waste materials in other countries. Installations have been reported in Tanzania,[10] Uganda,[11] and Bangladesh.[12] Since 1971 some experiments have also been carried out on the islands of the South Pacific; a pilot project was installed on a small farm in Fiji and a successful demonstration project was operated at Port Moresby, Papua New Guinea.[13] A prototype unit using chicken and turkey manure was recently established in Mexico.[14]

Finally, in the United States and Western Europe, interest in the use of anaerobic digestion to provide fuel and safe "natural" fertilizer for small-scale use has been growing steadily for a number of years and numerous pamphlets have appeared that give more-or-less detailed instructions for building digesters.[15]

Thus, extraction of energy from wastes by anaerobic digestion is decades old, and the general technology is well-known. It has not, however, been confined to small-scale use. Large-scale municipal digesters are used in the treatment of municipal sewage sludge, with the evolved gases satisfying part of the energy needs of the municipal treatment plant.

Common materials used for methane generation are often defined as "waste" materials, e.g., crop residues, animal wastes, and urban wastes including night soil. Some of these materials are already used in developing countries as fuels and/or fertilizers. Use of these materials for methane generation, as illustrated in Figure I-1, will allow additional value to be gained from them while the previous benefits are still retained.

Original Use:

```
        ANIMAL WASTES                              CROP RESIDUES

   applied directly to the    burned as fuel    plowed under to increase
   land as fertilizer                           soil organic matter and
                                                return nutrients to soil
                                HEAT
```

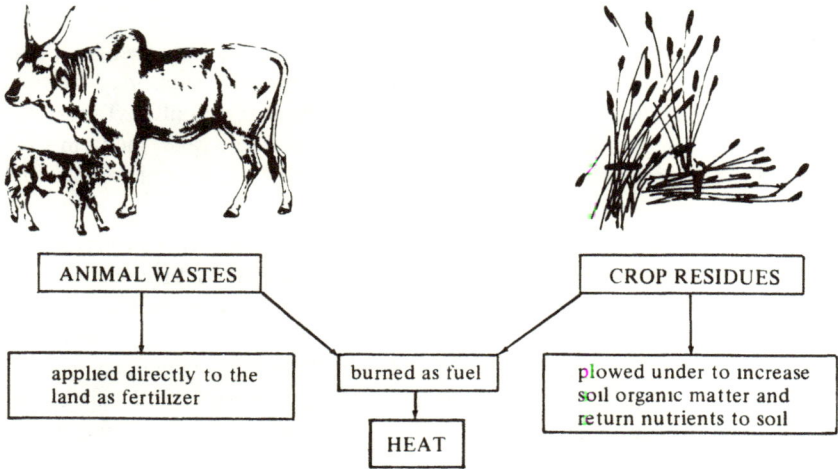

Addition of Methane Generation:

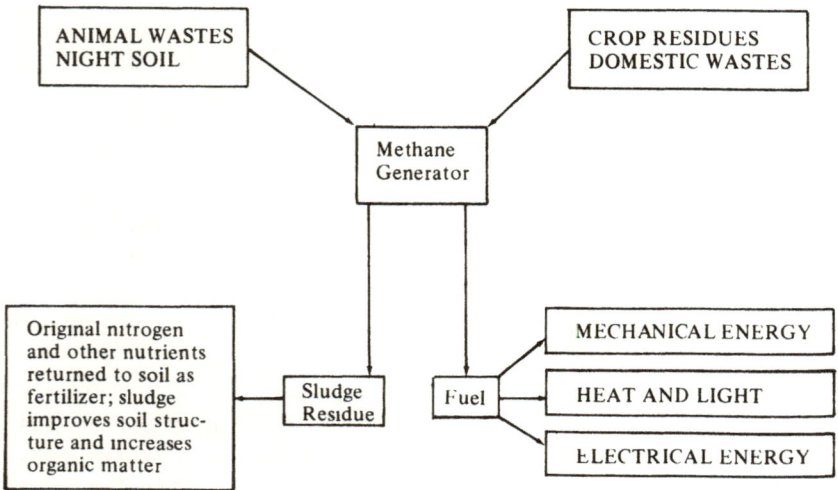

```
   ANIMAL WASTES                          CROP RESIDUES
   NIGHT SOIL                             DOMESTIC WASTES

                        Methane
                        Generator

   Original nitrogen                              MECHANICAL ENERGY
   and other nutrients
   returned to soil as          Sludge            HEAT AND LIGHT
   fertilizer; sludge           Residue    Fuel
   improves soil struc-                           ELECTRICAL ENERGY
   ture and increases
   organic matter
```

FIGURE I-1 Impact of anaerobic fermentation on use of organic wastes.

The use of rural wastes for methane generation, rather than directly as fuel or fertilizer, yields three direct benefits:

- The production of an energy resource that can be stored and used more efficiently;
- The creation of a stabilized residue (the sludge) that retains the fertilizer value of the original material; and
- The saving of the amount of energy required to produce an equivalent amount of nitrogen-containing fertilizer by synthetic processes.

Indirect benefits of methane generation include:

- Potential for partial sterilization of waste during fermentation, with the consequent reduction of the public health hazard of fecal pathogens; and
- Reduction, due to the fermentation process, of the transfer of fungal and other plant pathogens from one year's crop residue to the next year's crop.

References

1. McCabe, J., and Eckenfelder, Jr., W. W., eds. 1957. *Biological Treatment of Sewage and Industrial Wastes.* 2 vols. Vol. 2 *Anaerobic Digestion and Solids-Liquid Separation, Papers Presented at the Conference on Anaerobic Digestion and Solids Handling, Sponsored by Manhattan College, New York, April 24-26, 1957.* New York: Reinhold Publishing Co.
2. Lapp, H. M.; Schulte, D. D.; Sparling, A. B.; and Buchanan, L. C. 1975. Methane production from animal wastes. 1. Fundamental considerations. *Canadian Agricultural Engineering* 17(2):97-102.
3. At that time, the Imperial Agricultural Research Institute.
4. Idnani, M. A., and Acharya, C. N. 1963. *Bio-Gas Plants.* Farm Bulletin No. 1 (New Series). New Delhi: Indian Council of Agricultural Research.
5. Yu, Ju-tung. 1965. *Notes on Raising Pigs to Gain Wealth.* Taipei: Feng Nien She.
6. Chung, Po. 1973. Production and use of methane from animal wastes in Taiwan. In *Proceedings, International Biomass Energy Conference, 13-15 May 1973, Winnipeg, Manitoba, Canada.* Winnipeg: Biomass Energy Institute (P.O. Box 129, Winnipeg, Manitoba R3M 3S7, Canada).
7. *BBC Summary of World Broadcasts,* FE/W698/A/15, 8 Nov. 1972.
8. *BBC Summary of World Broadcasts,* FE/W770/A/10, 10 April 1974.
9. Institute of Agricultural Engineering and Utilization. Office of Rural Development. 1973. *Present Status of Methane Gas Utilization as a Rural Fuel in Korea.* Suwon, Korea: Institute of Agricultural Engineering and Utilization. Office of Rural Development.
10. Robson, John R. K. 1975. Personal communication.
11. Jeffries, Sir Charles, ed. 1964. *A Review of Colonial Research, 1940-1960.* London: Her Majesty's Stationery Office.
12. Chan, George L. 1972. Waste Utilization in Rural Industrialization, with Particular Reference to the South Pacific. Sixth Waigani Seminar, Priorities in Melanesian Development, 6-20 May 1972, University of Papua New Guinea, Port Moresby, Papua New Guinea.

13. Jedlicka, Allen D. 1974. Comments on the Introduction of Methane (Bio-gas) Gener-
    ation in Mexico with an Emphasis on the Diffusion of Back-yard Generators for Use
    by Peasant Farmers. Cedar Falls, Iowa: University of Northern Iowa, College of
    Business and Behavioral Sciences. (Unpublished paper.)
14. For example: Fry, L. J., and Merrill, R. 1973. *Methane Digesters for Fuel, Gas and
    Fertilizer.* Newsletter No. 3. Santa Cruz, California: New Alchemy Institute; Now. . .
    electricity from manure gases. 1970. *The Mother Earth News* 3(May):44. reprinted
    by special permission from *Farm Journal,* May 1963; Gilbert, Keith D. 1970. How
    to generate power from garbage. *The Mother Earth News* 3(May):45-53; Sampson,
    Steven. 1975. *Methane (Atomic Rooster's Here).* Wadebridge, Cornwall, England:
    Wadebridge Ecological Centre.
15. Garg, A. C.; Idnani, M. A.; and Abraham, T. P. 1971. *Organic Manures.* Bulletin
    (Agric.) No. 32. New Delhi. Indian Council of Agricultural Research.

Chapter 2

System Description

The biogas system includes the gas production process, the use of the gas produced, and the use of the sludge remaining after fermentation is complete.

Gas Production and Use

The production of methane during the anaerobic digestion of biologically degradable organic matter depends on the amount and kind of material added to the system. The efficiency of production of methane depends, to some extent, on the continuous operation of the system. As much as 8–9 ft^3 of gas (containing 50–70 percent methane) can be produced per pound of volatile solids* added to the digester (0.5–0.6 m^3/kg) when the organic matter is highly biodegradable (e.g., night soil or poultry, pig, or beef-cattle fecal matter).** Combustion of 1 ft^3 (about 30 l) of gas will release an amount of energy equivalent to lighting a 25-watt bulb for about 6 hours.

In general, lower gas-production rates result when the wastes are less biodegradable. In developing countries, an important consideration will be the differences in the quantity and quality of waste material produced from various sources; for example, the quality and quantity of animal manure is influenced by the diet and general health of the animals. Table I-1 lists potential sources of organic matter for methane generation.[1]

The use to which the gas is put depends upon removal of noncombustible components (such as carbon dioxide) and corrosive components (such as hydrogen sulfide). Among the many potential uses of digester gas are hot-water heating, building heating, room lighting, and home cooking. Gas from a digester can be used in gas-burning appliances if they are modified for its use. Conversion of internal-combustion engines to run on digester gas can be relatively simple; thus the gas could also be used for pumping water for irrigation. Past experience has shown that where methane is generated in

*See Glossary.
**See footnote Part III, Chapter 1, p. 63.

TABLE I-1   Organic Matter with Potential for Methane Generation[1]

Crop Wastes	Sugar cane trash, weeds, corn and related crop stubble, straw, spoiled fodder
Wastes of Animal Origin	Cattle-shed wastes (dung, urine, litter), poultry litter, sheep and goat droppings, slaughterhouse wastes (blood, meat), fishery wastes, leather, wool wastes
Wastes of Human Origin	Feces, urine, refuse
By-products and Wastes from Agriculture-Based Industries	Oil cakes, bagasse, rice bran, tobacco wastes and seeds, wastes from fruit and vegetable processing, press-mud from sugar factories, tea waste, cotton dust from textile industries
Forest Litter	Twigs, bark, branches, leaves
Wastes from Aquatic Growth	Marine algae, seaweeds, water hyacinths

significant quantities in rural areas of developing countries, its use is primarily for lighting and cooking.

The gas produced by digestion of organic waste is colorless, flammable, and generally contains approximately 60 percent methane and 40 percent carbon dioxide, with small amounts of other gases such as hydrogen, nitrogen, and hydrogen sulfide. It has a calorific value of more than 500 Btu/ft^3 (18,676 kJ/m^3). Methane itself is a nontoxic gas and possesses a slight but not unpleasant smell; however, if the conditions of digestion produce a significant quantity of hydrogen sulfide, the gas will have a distinctly unpleasant odor.

Uses of Sludge

The organic fraction of sludges from an anaerobic digester operating on plant and animal waste may contain up to 30-40 percent of lignin and undigested cellulose and lipid materials, on a dry-weight basis, depending on the type of raw material used. The remainder consists of substances originally present in the raw material but protected from bacterial decomposition by lignin and cutin, newly synthesized bacterial cellular substances, and relatively small amounts of volatile fatty acids. The amount of bacterial cell mass is small (less than 10-20 percent of the substrate is converted to cells). Therefore, there is less risk of creating odor and insect-breeding problems when anaerobically digested sludges are stored or spread on land than there is when untreated or partially treated organic waste materials are similarly handled or are indiscriminately disposed of or stored. (See Part II, Chapter 4, Residue: Composition, Storage, and Use.)

One of the direct benefits of the anaerobic process mentioned earlier is that the nutrient elements in the plant residues and animal wastes used as raw materials are conserved for the production of subsequent crops. Among these nutrients is nitrogen, practically all of which is conserved. Since it is often present in the sludge in the form of ammonia, proper storage of sludge and application to the land is needed to minimize the loss of this volatile chemical. (See Part II, Chapter 4.) All other chemical elements (except carbon, oxygen, hydrogen, and some sulfur) contained in plant residues and animal wastes are conserved in anaerobically digested sludge.

The end result of applying anaerobically digested sludge to soils has the same effect as that obtained from applying any other kind of organic matter. The humus materials formed improve physical properties of soil: for example, aeration, moisture-holding capacity, and water-infiltration capacity are improved and cation-exchange capacity is increased. Furthermore, the sludge serves as a source of energy and nutrients for the development of microbial populations that, directly and indirectly, improve the solubility, and thus the availability to higher plants, of essential chemical nutrients contained in soil minerals.

Experience has shown that anaerobically digested sludge from municipal wastewater treatment plants receiving large loads of industrial waste have not caused conditions toxic to plants even in heavy applications to agricultural lands in industrialized countries.[2] Thus the possibility is remote that heavy metals or pesticides contained in digested plant and animal waste would cause any problem with digester sludge applied to soils in developing countries. Using the sludge as fertilizer may result in enhanced levels of concentration of some elements in plant tissues, but from the standpoint of animal and human nutrition this is more likely to be a benefit than a detriment. For the most part, the elements likely to be increased in plant tissues are essential animal nutrients.

Elements of a Successful System

The main requisites for successful operation of a methane-generation system are acceptance by potential users, ability to use the gas when produced, sufficient demand for the gas, availability of sufficient raw material to meet the production requirement, and adequate maintenance and operational control.

Many scientists and engineers with experience in the anaerobic-digestion process are cautiously optimistic about the prospects of extracting energy from organic wastes. The fundamentals of the process are well known and there is a significant quantity of unused organics in rural areas. These organics will produce large quantities of methane gas. The present technology can be utilized and adapted to local conditions where necessary, but this must be done with competent guidance.

In sum, then, before the anaerobic digester concept is adopted in any country or applied to specific wastes, the panel recommends that the following conditions obtain:

- The equipment must be demonstrated to be functional at the scale of the proposed operation;
- The installation should be accompanied by clear instructions—including written instructions where practical—for operation;
- The equipment and its capacity must be suitable for the quantities and types of material to be handled and compatible with other components; and
- Users and/or operators must be capable of properly maintaining and operating the equipment.

Adequate methane production requires proper digester conditions. Maximum gas generation per unit of input material in the shortest time depends on obtaining and maintaining optimal conditions for bacterial activity and anaerobic fermentation. The critical factors are outlined in Table I-2.

Digester design, and output expectations, must be tailored to the resources, climatic conditions, and building materials found in each locale; thus a technical assistance program may be required in each region of the country in which methane generation is feasible. To minimize capital outlay for equipment, digesters should be of a size suitable for local demand and, whenever possible, should be constructed of local materials available close to the site of operation. Fabrication from corrosion-resistant materials such as wood, ferrocement, concrete, brick, or stone, rather than metal, may also reduce costs by extending equipment life.

TABLE I-2   Critical Factors Affecting Methane Production

Initiating Digester Operation	Anaerobic conditions
	Appropriate substrates
	Appropriate bacterial type
	Appropriate environmental conditions in the digester
Maintaining Output of a Functioning Digester	Protection of digester from sudden environmental changes—maintains adequate population of the anaerobic bacteria
	Steady supply of substrate—continuous operation ensures a higher output than intermittent use; semicontinuous operation is more practical on a small scale
	Removal of inert wastes such as sand and rocks—prevents wear on mechanical parts
	Digester temperature—heating the digester will increase the rate of bacterial activity within the mesophilic (30°-40°C) and thermophilic (50°-60°C) temperature ranges, thus increasing methane yield and decreasing the detention time of the substrate in the digester

Some technical assistance may also be required to determine startup conditions for each region of the country.

In addition, once the digesters are functioning, there may be need for
occasional technical assistance. Digester operation can be relatively simple,
however, once a knowledge of construction materials, available organic matter, energy requirements, and user technical expertise have been acquired.

The criteria for determining the design parameters for a methane-generating system are illustrated in Figure I-2 and outlined below:

1. *Determine the production requirements of biogas.* The size of a methane unit depends on the quantity of gas needed. The total production requirements can be determined by itemizing the applications for which the methane
will be used and summing up the quantities of gas needed for each use.

2. *Inventory the raw waste materials.* This should include animal, agricultural, human, and any other wastes that can be digested for the production
of methane.

3. *Determine the time needed to accomplish the optimal digestion of the
waste materials ("detention time").* The detention time will be shorter if the
digester is heated. In the case of the small digesters (producing less than 500
ft^3 [14 m^3] of gas per day), it may not be advisable to incorporate heating
with fuel because of the maintenance required. It may, however, be advisable
to consider solar heating of the digester contents.[4]

4. *Determine the size of the digester.* The minimum volume of the digester can be determined by multiplying the detention time by the volume of
material that must be added each day to produce the desired volume of gas
daily.

5. *Decide whether to compartmentalize and divide the digester volume
into two or more stages.* In plants intended to serve an entire village or community, a digester with large capacity is usually needed, and adequate mixing
may be difficult to achieve in a large undivided digester.

6. *Determine the size of the gas holder.* The volume of the gas holder
depends upon daily production and usage and may be as low as 50 percent of
the total volume of gas produced daily, provided the gas is used frequently. It
should be larger if gas is used at irregular intervals.

Safety Concerns

Safety concerns related to methane generation include health hazards and
risks of fire or explosion. Since methane gas is flammable and can be explosive when mixed with air, suitable safety instructions should be provided
when the digester is installed.

A minor impurity of the biogas produced by anaerobic fermentation is
hydrogen sulfide, a toxic gas that is highly corrosive in water solution. However,

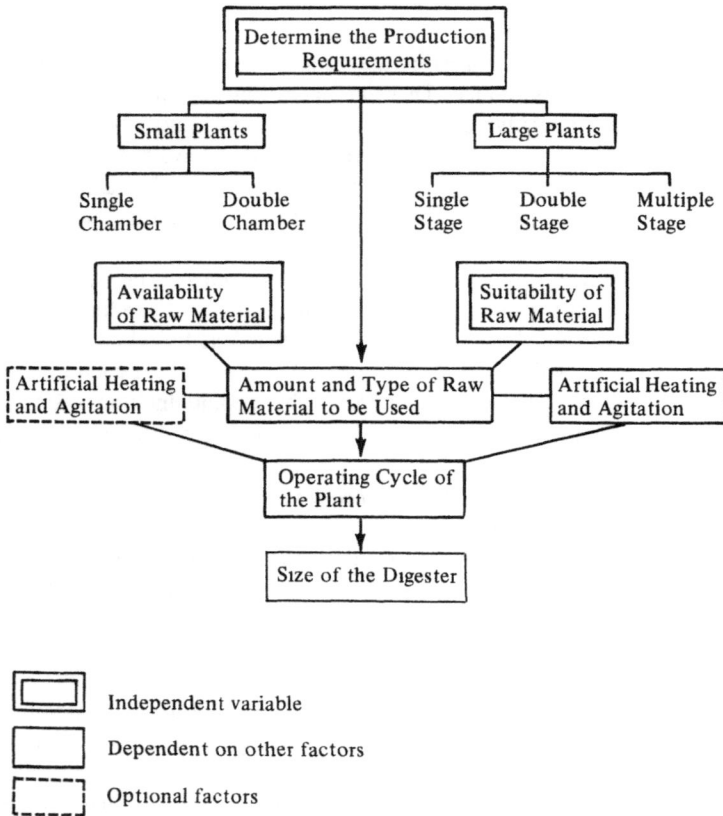

FIGURE I-2 Design parameters for methane-generating systems. (Adapted from Singh. Ref. 3)

hydrogen sulfide can be safely removed from the digester gas by a variety of methods, including bubbling the gas through lime water.

On the other side of the safety coin are advantages inherent in this process: properly operated methane generating units reduce not only the public health hazard of fecal pathogens but the transfer of plant pathogens between successive crops, and provide a valuable, nutrient-rich fertilizer.

Table I-3 summarizes some of the advantages and disadvantages of the use of anaerobic digestion as a method of processing biodegradable organic waste materials.

TABLE I-3   Advantages and Disadvantages of Anaerobic Digestion

Advantages	Disadvantages
Produces large amount of methane gas. Methane can be stored at ambient temperature.	Possibility of explosion.
Produces free-flowing, thick, liquid sludge.	High capital cost. (However, if operated and maintained properly, the system may pay for itself.)
Sludges are almost odorless, odor not disagreeable.	May develop a volume of waste material much larger than the original material, since water is added to substrate. (This may not be a disadvantage in the rural areas of developing countries where farm fields are located close to the village, thus permitting the liquid sludge to be applied directly to the land, serving both for irrigation and as fertilizer.)
Sludge has good fertilizer value and can be used as a soil conditioner.	
Reduces organic content of waste materials by 30–50 percent and produces a stabilized sludge for ultimate disposal.	
Weed seeds are destroyed and pathogens are either destroyed or greatly reduced in number.	Liquid sludge presents a potential water-pollution problem if handled incorrectly.
Rodents and flies are not attracted to the end product of the process. Access of pests and vermin to wastes is limited.	Maintenance and control are required.
Provides a sanitary way for disposal of human and animal wastes.	Certain chemicals in the waste, if excessive, have the potential to interfere with digester performance. (However, these chemicals are encountered only in sludges from industrial wastewaters and therefore are not likely to be a problem in a rural village system.)
Helps conserve scarce local energy resources such as wood.	
	Proper operating conditions must be maintained in the digester for maximum gas production.
	Most efficient use of methane as a fuel requires removal of impurities such as CO_2 and H_2S, particularly when the gas is to be used in internal-combustion engines.

References

1. Garg, A. C.; Idnani, M. A.; and Abraham, T. P. 1971. *Organic Manures.* Bulletin (Agric.) No. 32. New Delhi: Indian Council of Agricultural Research.
2. Hinesly, T. D.; Jones, R. L.; Ryler, J. J.; and Ziegler, E. L. 1976. Soybean yield responses and assimilation of Zn and Cd from sewage sludge-amended soil. *Journal of the Water Pollution Control Federation* 48:2137-2152.
3. Singh, Ram Bux. 1971. *Bio-gas Plant: Generating Methane from Organic Wastes,* p. 36. Ajitmal, Etawah (U.P.), India: Gobar Gas Research Station.
4. National Academy of Sciences. Commission on International Realtions. 1976. *Energy for Rural Development: Renewable Resources and Alternative Technologies for Developing Countries.* Report of an *ad hoc* panel of the Advisory Committee on Technology Innovation. Board on Science and Technology for International Development. Washington, D.C.: National Academy of Sciences.

Chapter 3

Economic Feasibility
of Methane Production

Although anaerobic digester technology has been used successfully in some developing countries, economic considerations may preclude or limit its adoption in others. The economic feasibility of methane generation from wastes may vary widely. It is dependent on factors that include the availability of domestic sources of energy, the cost of imported fuel, the uses and actual benefits from methane production, public and private costs associated with the development and utilization of methane, and on the technology used to generate methane. Furthermore, energy is required to process methane and this must be considered when evaluating the amount of energy produced. All these factors must be taken into account in any benefit/cost analysis, along with an assessment of the skills available for a successful project.

Domestic Energy Sources

The existence of other domestic energy sources, with which locally produced methane might compete as a fuel, must be considered in judging the feasibility of a methane-production program. In making this judgment, however, one must consider not only the availability of such energy sources, but their economic accessibility to the domestic user, particularly in rural areas. Undeveloped coal and petroleum reserves may have little relevance for developing countries, particularly for rural areas, at least for a long period. The opportunity costs of coal, wood, and other basic energy sources must also be taken into account; they may be quite high for direct family use. For example, in Korea there is a shortage of wood and wood conservation is a priority in national planning programs.[1]

Raw Materials

The quantities and nature of raw materials used for methane generation vary widely from country to country and within a particular country. Hence, the technology for collecting and processing raw materials may vary, and different economic problems may be encountered. Raw materials may be obtained from a variety of sources, including wastes from livestock and poultry, humans, crops, and food processing. Different problems may be encountered with each of these wastes with regard to collection, transportation, processing, storage, residue disposal, and eventual use. For a given waste, such as animal manure, the economic feasibility of gas production may also be greatly influenced by qualitative differences stemming from the conditions under which the waste is produced, such as the different rations fed to various types of animals.

Production and Utilization Processes

Methane production and utilization involves collection, transportation, processing, storage of product and residue, and utilization processes, as illustrated in the following flow diagram.

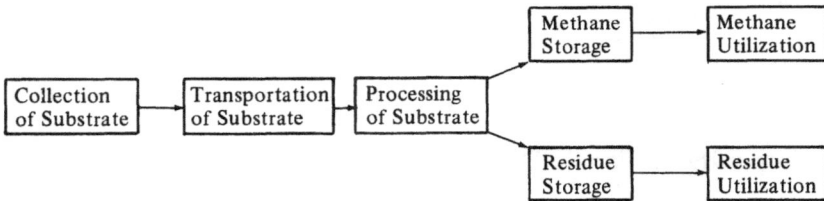

A feasibility study for methane production must include the entire production and utilization system, involving all these processes; capital and labor requirements and annual costs must be determined for all processes and must be related to local material, labor, and price conditions. Materials not locally available may have to be excluded in the technological design. Also included in a cost analysis should be the cost of removing any impurities from the resultant gases. The major cost factors to be considered are summarized in Table I-4. The technical design should, if possible, make use only of construction materials that are locally available.

The various ways in which methane can be used have already been mentioned. The actual uses of methane in a specific country may be limited; for example, the amount of gas produced, if small, would preclude some uses. Thus, the actual use to be made of the methane must be clearly specified in any anaerobic digester feasibility study.

TABLE I-4   Major Cost Factors in Methane Production and Utilization Processes

Costs Associated with Processing	Public Costs	Costs to User
1. Installation costs 　　Capital 　　Labor	1. Demonstration facilities	1. Share of installation
	2. Sharing arrangements 　　for installation costs	2. Annual costs
2. Annual costs 　　Maintenance 　　Labor	3. Technical assistance	

Gas Utilization Costs	Collection and Transportation Costs	Residue Costs
1. Storage	1. Labor	1. Labor
2. Distribution system	2. Equipment for transport	2. Transportation to storage site
3. Adaptation of existing equipment to use of methane as fuel	3. Raw material (if purchased)	3. Storage
4. Purchase of additional equipment adapted to use of methane as fuel		4. Transportation to utilization site

Community vs. Family Digesters

Most methane production and use in developing countries has been from family digesters. Community digesters are also feasible in many situations, but resource requirements differ from those for the smaller family digesters. Community digesters require a larger supply of raw materials, and perhaps a different technological design. They also need higher managerial and technical skills for operating the equipment because of the problems of materials handling and product and residue storage and use that accompany larger-scale operation. Although limited information is available on possible economies of size for digesters in developing countries, published reports argue in favor of community (or institutional) digesters for a variety of social reasons. [2,3]

The design criteria for a methane generator should be based on minimum-size considerations. For a digester to be economically worthwhile, a minimum amount of both raw materials and methane output for a specific use must be available. The minimum-size digester may vary with many factors, including the use that is made of methane, the types of raw materials used, and the technological design.

Managerial and Technical Skills

Both public and private managerial and technical skills will be required in most countries to develop a successful methane program. Technical assistance will be required to train local people for the installation, operation, and maintenance of methane digesters, and, perhaps, to assure continued use after digesters have been adopted by local people. The importance of technical assistance provided by the government should not be minimized.

Benefit/Cost Appraisal

The preceding considerations should be reflected in any benefit/cost analysis of a development project for methane production. Careful consideration should be given to selection of the time period for evaluating the stream of benefits, the discount rate used (opportunity cost of capital), and the kinds of benefits to include and their evaluation, including the reflection of industry supply effects. Several benefit/cost appraisals should be made to reflect alternative assumptions with regard to these factors.

In evaluating the benefits of methane generation in a developing country, the planners should carefully identify and delineate the possible uses noted above. Planners should be careful, however, to enumerate only those benefits *likely* to accrue from a particular public project. The direct benefits include those accruing from heating, cooking, lighting, refrigeration, or creation of fuel for internal-combustion engines such as those used to drive water pumps. However, other benefits that may also accrue from this technology, such as improved public health, agricultural productivity increases, release of other fuels for other uses, diminished pressure on forest reserves because of reduced demand for firewood, and increased employment, must also be taken into account.

In order for a methane-generation program to be successful in a developing country, some initial financial and continuing technical assistance from central and local governments may be needed. Part of the materials cost of individual methane generators might have to be borne by central and/or local governments, and trained technical personnel would be needed to help local people learn to install, operate, and maintain methane digesters. Demonstration facilities may be required, with early adopters subsidized for demonstration purposes to encourage the large-scale adoption of this technology. The particular educational means needed to induce local people to adopt methane digesters may vary from country to country and between regions within a country. The public costs, therefore, will be subject to some variation among countries.

Although the benefit/cost ratio may be favorable for methane-generation development, as with other proposed investments the opportunity costs of public capital must always be considered. Capital for land development and water control or other public projects may be more crucial in some countries than the development of methane production. This is not to say that methane production may not be a feasible development project. Other projects may simply have a greater urgency or promise a higher return.

Benefit/Cost Analysis

A Korean study of methane generation illustrates the use of benefit/cost analysis.[1,4] Several types of beneficial impacts were identified. The study showed that the use of methane as a fuel substitute (primarily for cooking) would conserve other domestic energy sources, such as rice and barley straw, wood, coal, and oil. Wood is in extremely short supply in Korea (see Ref. 1),

TABLE I-5   Item Cost of Generator Installation, Korea, 1969-70[a]

Item	Quantity	Cost (Won)
Cement	15 bags	4,500
Wooden Materials		8,270
Vinyl Holder		4,000
Clay Pipe	4 each	1,200
Nails and Tar		774
Vinyl Hose and Burner	1 set	1,850
Other Materials		4,400
Labor		2,600
Total		27,594

[a]Source: Kim and Libby (1972).[4] It was estimated that an annual maintenance cost of 4,000 won would be required. (In 1969-70, 270 W = US $1.00.) Presumably the figures refer to a 100-ft^3 digester, although this is not specified.

TABLE I-6   Estimated Economic Benefit from Use of Methane for Cooking, Korea, 1970[a]

Factor	Quantity Saved (Thousand Metric Tons)	Economic Value (Million Won)
Rice and Barley Straw	2,799	9,900
Chemical Fertilizer	130	2,600
Yield Increase in Rice	70.1	4,800
Wood	3,317	12,300
Coal	29,750	3,000
Oil	15,000	
Total		32,600

[a]Source: Kim and Libby (1972).

and straw has other uses, such as in the production of bags or as animal feed. Methane production also provided a valuable fertilizer, safer because the digested material was more parasite- and disease-free than the raw wastes. Public health could be enhanced, and a cheaper, more efficient fertilizer could be produced than was otherwise available. Labor-saving benefits could also be realized. With labor supply critical in Korea at certain peak periods during transplanting and harvest, the use of methane could reduce to 1 hour the average 3 or 4 hours spent each day gathering fuel and cooking for a family of four. The estimated aggregate economic benefit from the use of methane in Korea in 1970 is shown in Table I-6.*

The installation and annual maintenance costs for a typical methane generator in Korea were estimated to be 27,594 and 4,000 won, respectively (Table I-5). The Korean government provided technical and financial assistance for the methane development project, but the aggregate benefits, as outlined in Table I-6, appeared to justify the public costs of providing such assistance for the number of generators installed.

Other cost estimates have been made. A number of Indian studies provide generally favorable economic evaluations of biogas based on cost estimates for methane production and use.[5] Cost estimates are also available in a recent study of small-scale methane generators in Mexico, but these are based on small experimental prototypes that have not yet been built on a practical single-family scale.[6]

Other case studies have demonstrated not only significant economic benefit, but also some intangible social benefits that have resulted from the installation of methane-generation (biogas) systems.[7] The results of an economic analysis of 10 installations in India are shown in Table I-7. Although not spelled out in detail in this study, the economic benefits of the fertilizer produced were obviously significant. The study showed that the amount of fertilizer available from the biogas plants was about 60 percent greater than that formerly available to the farmers when dung was burned for fuel. Consequently, crop productivity increases up to 75 percent were reported, along with a reduction in the amount of chemical fertilizers used.

Among the other benefits reported in this study were improved household appearance and cleanliness as a result of the lack of soot production from dung or wood cooking fires; more time available for the women for other tasks such as spinning cotton; and an improvement in family health.

---

*The types and amount of benefits realized in this study would not necessarily be comparable to those that might be realized in other developing countries; other benefits might accrue, depending on the ultimate use made of methane. For example, methane could be used for other purposes, such as electricity generation or for power for pumping irrigation water. In countries where employment for technical graduates is a problem, they might be used to provide the needed technical assistance to the local populace.

TABLE I-7 Details Showing Economics of the Plants in the Sewapuri Area [Uttar Pradesh, India] (a)

(1) Plant	(2) Total investment of plan (Rs.) (b)	(3) Number of persons who take meal	(4) Annual cost of fuel if biogas were not used (Rs)	(5) Present annual comsumption Fuel (Rs.)	(6) Present annual comsumption Gas ft³	(7) Annual capital charges (depreciation) (Rs.)	(8) Annual labor charges for slurry (Rs.)	(9) Annual maintenance and other charges (Rs.)	(10) Interest charges (8 percent on capital) (c)	(11) Annual cost of production (Rs.)	(12) Annual net savings (Rs)	(13) Net return on the cost after allowing operational expenditure (percent)	(14) Cost of gas per 1,000 ft³ (Rs)
1	2,400	30	1,080	—	134,250	80	60	24	192	356	724	30.2	2.65
2 (d)	—	—	—	—	—	—	—	—	—	—	—	—	—
3	1,500	5	310	—	52,195	50	60	15	120	245	65	4.3	4.69
4	5,000	40	1,600	—	182,500	167	120	50	400	737	863	17 3	4.03
5	1,850	16	600	—	73,000	62	72	19	148	301	299	16.2	4 12
6	1,890	25	1,000	—	91,250	60	60	19	144	283	717	39 8	3 10
7	1,900	7	360	—	45,625	63	60	19	152	294	66	3.5	6 44
8	1,500	20	620	—	54,750	50	60	15	120	245	375	25.0	4.47
9	2,000	16	564	—	91,250	67	60	20	160	307	257	12.9	3.36
10	7,000	125	3,160	600	202,940	233	500	70	560	1,363	1,797	25 7	6.72

(a) Adapted from Prasad (1967). [7] Costs are given in rupees: current rate is approximately Rs. 8.0 = US $1.00.
(b) The total cost of each plant includes the cost of the digester, stoves, pipe fittings and the costs of labor involved in the installation.
(c) Current interest rates in India are higher. The state of Uttar Pradesh now charges 17 percent interest on biogas plant construction loans.
(d) No figures available.

22

Other cost/benefit figures from the Indian experience are given in Appendix 2.

References

1. Exacerbating the problem of fuel in rural areas is the growing shortage of firewood, a shortage that is reaching critical proportions in many countries. This problem is discussed in: Eckholm, Erik P. 1975. *The Energy Crisis: Firewood*. Worldwatch Paper 1. Washington, D.C.: Worldwatch Institute.
2. Sathianathan, M. A. 1975. Bio-gas plants in use. In *Bio-gas Achievements and Challenges*, Chapter 12. New Delhi: Association of Voluntary Agencies for Rural Development.
3. Mardon, C. J. 1976. An Assessment of Small-Scale Methane Gas Production and Methane-Algal Systems in the Asian-Pacific Area. Report to CSIRO, Division of Chemical Technology, Melbourne, Australia.
4. Kim, S. G., and Libby, L. W. 1972. Rural infrastructure. In *Korean Agricultural Sector Study Special Report No. 2*, pp. 69-75. East Lansing: Michigan State University, Department of Agricultural Economics.
5. Prasad, C. R.; Prasad, K. Krishna; and Reddy, A. K. N. 1974. Biogas plants: prospects, problems and tasks. *Economic and Political Weekly* (India) 9:1347-1364.
6. Jedlicka, Allen D. 1974. Comments on the Introduction of Methane (Bio-gas) Generation in Mexico with an Emphasis on the Diffusion of Back-yard Generators for Use by Peasant Farmers. Cedar Falls, Iowa: University of Northern Iowa, College of Business and Behavioral Sciences. (Unpublished paper.)
7. Prasad, Hanuman. 1967. Gobar gas plants: an empirical study. *Khadi Gramodyog: Journal of Rural Economy* (India) 13:677-688.

PART   II

TECHNOLOGY OF ANAEROBIC FERMENTATION: STATE OF THE ART

Chapter 1

Biological Mechanisms

Anaerobic treatment of complex polymeric organic fibers may be considered to be a three-stage process, as shown in Figure II-1. In the first stage, a group of facultative microorganisms acts upon the organic substrates. By enzymatic hydrolysis the polymers are converted into soluble monomers that become the substrates for the microorganisms in the second stage, in which the soluble organic compounds are converted into organic acids. These soluble organic acids—primarily acetic acid—are the substrate for the final stage of decomposition accomplished by the methanogenic bacteria. These bacteria are strictly anaerobic and can produce methane in two ways: by fermenting acetic acid to methane and carbon dioxide, or by reducing carbon dioxide to methane using hydrogen gas or formate produced by other bacteria. The production of methane, a gas, in the third stage reduces the amount of oxygen-demanding material remaining. This produces a biologically stable residue.

Bacterial growth occurs during all stages of the fermentation process; the proportion of total substrate utilized to support bacterial growth, however, is low compared to that utilized in aerobic biological processes.

Substrates and Microbiology of State 1: *Polymer Breakdown*

The initial substrates for stage 1 will be various waste materials (see Table I-1) composed primarily of carbohydrates, with some lipid, protein, and inorganic material. The major carbohydrates are cellulose and other components of plant fiber, such as hemicellulose and lignin. These are found not only in crop residues but also in animal wastes, since they are often not digestible. A broad spectrum of anaerobic bacteria is required to solubilize these materials, including bacteria possessing cellulolytic, lipolytic, and proteolytic enzymatic capacity.

Cellulolytic activity is the most critical in reducing the complex raw material to simple, soluble, organic components. The largest fraction of the organic matter in sewage sludge is cellulose (Table II-1), and if crop residues

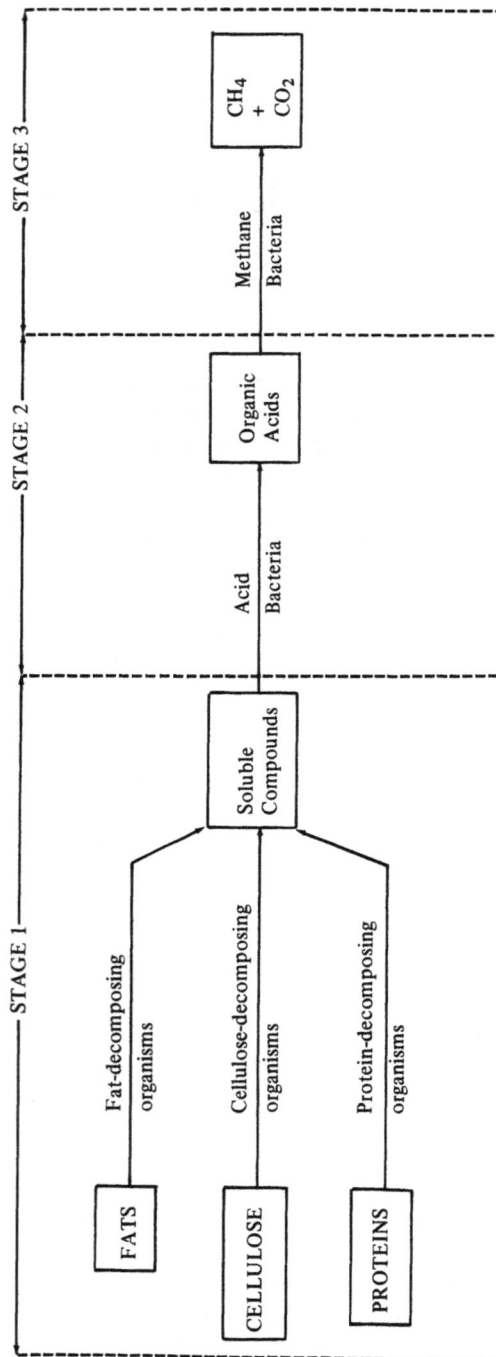

FIGURE II-1 Anaerobic fermentation of organic solids.

28

TABLE II-1   Chemical Components of a Sewage Sludge[a]

Component	Percent (dry weight basis)
Hemicellulose	6.0
Cellulose	34.5
Lipids	14.0
Protein	19.0
Ash	34.0

[a]Source: Maki (1954).[1]

are utilized directly, an even higher proportion of the total dry matter will be cellulose. Cellulose consists of polymerized glucose units in a chain of indefinite length with complex branching patterns. Cellulolytic bacteria sequentially reduce the chains and branches to dimeric and then to monomeric sugar molecules, which are then converted to organic acids.

The cellulolytic bacteria are usually divided into two classes on the basis of the optimal temperature at which digestion occurs. Mesophilic bacteria have optima in the range of 30°-40°C, as in the rumen of cattle, while thermophilic species work optimally at 50°-60°C. Both groups have pH optima in the range of 6.0-7.0. As organic acids are produced during the breakdown of cellulose, the pH may fall; during the initiation of the fermentation and during the digestive process, it may be necessary to buffer the system with lime to stabilize it. When the acid-forming bacteria of stage 2 and the methanogenic bacteria of stage 3 are present in a balanced reaction, the pH of the entire system will reach an equilibrium value of about 7, since the organic acids will be removed as they are produced.

The synergistic (cooperative) action of a variety of cellulolytic and other hydrolytic bacterial species is important in the breakdown of the raw materials. Studies have shown that the rate of cellulose removal by mixed cultures is substantially greater than the rate of cellulose removal by most cellulolytic bacteria maintained as pure cultures. This implies that there is a synergistic action involved as cellulolytic bacteria by-products are removed by non-cellulolytic bacteria. A noticeable lag in the initiation of gas production also suggests that the establishment of the proper flora may be the rate-limiting step in cellulose decomposition.

The conversion of cellulose and other complex raw materials to simple monomers is probably the rate-limiting step in methane production, since bacterial action is much slower in stage 1 than in either stage 2 or 3. The hydrolysis rates under several different sets of conditions of temperature, substrate, and bacterial species are shown in Table II-2. Hydrolysis rate is dependent on substrate and bacterial concentration as well as on the environmental factors of pH and temperature.

TABLE II-2 Summary of Hydrolysis Rate of Cellulose in Anaerobic Fermentation

Reference	System and Culture	Initial Cellulose Concentration (mg/1)	Cellulose Material	pH	Hydrolysis Rate (mg/1 per day)
Maki[1]	Batch, mixed 2 pure cultures from sewage, 38°C, mesophilic	2,000	Whatman #1 filter paper	6.8	(1)* 260 (2)* 660
Heukelekian[2]	Batch, pure culture from sewage, 25°C, mesophilic	3,120	Cellulose in sewage sludge	7.4	142
McBee[3]	Batch, pure culture from soil and manure, 55°C, thermophilic	(1)* 744 (2)* 2,980	Absorbent cotton	– –	(1)* 149 (2)* 426
Stranks[4]	Batch, mixed culture from rumen, 60°C, thermophilic	41,200	Whatman #2 filter paper	6.5	11,400

*Indicates experiments with different strains.

Substrates and Microbiology of Stage 2: *Acid Production*

The monomeric components released by the hydrolytic breakdown that occurs during stage 1 bacterial action become the substrate for the acid-producing bacteria of stage 2. The acids are produced as the end products of bacterial metabolism of carbohydrate; acetic, proprionic, and lactic acids are the major products. Methanogenic bacteria are very restricted in substrate utilization and are probably capable of utilizing only the acetic acid. Some species of methanogenic bacteria can produce methane from hydrogen gas and carbon dioxide; these substrates are also produced during carbohydrate catabolism. Methane can also be produced by the reduction of methanol, another possible by-product of carbohydrate breakdown. However, acetic acid is probably the single most important substrate for methane formation; studies following the fate of various substrates indicate that approximately 70 percent of the methane produced was from acetic acid.[5]

The microbiology of the stage 2 processes is not well understood. Many bacterial species are involved, and the proportion of acids, hydrogen gas, carbon dioxide, and simple alcohols produced depends on the flora present as well as on the environmental conditions.

Substrates and Microbiology of Stage 3: *Methane Production*

Methanogenic bacteria, as pointed out above, are highly restricted in substrate utilization and probably utilize only those substrates mentioned above that are produced by other bacteria during stage 2. Methanogenic bacteria utilize acetic acid, methanol, or carbon dioxide and hydrogen gas to produce methane. A few other substrates can be utilized, such as formic acid, but these are not important, since they are not usually present in anaerobic fermentations. Methanogenic bacteria are also dependent on the stage 1 and 2 bacteria to provide nutrients in a useful form; for example, organic nitrogen compounds must be reduced to ammonia to ensure efficient nitrogen utilization by the methanogenic bacteria. These bacteria also require phosphate and other materials; their exact requirements, however, have not been determined.

Methane bacteria are also sensitive to certain environmental factors. Because they are obligate anaerobes, their growth is inhibited even by small amounts of oxygen and it is essential that a highly reducing environment be maintained to promote their growth. Not only oxygen, but any highly oxidized material, such as nitrites or nitrates, can inhibit methanogenic bacteria.

These bacteria are also very sensitive to changes in pH; the optimal pH range for methane production is between 7.0 and 7.2, although gas production is satisfactory between 6.6 and 7.6. When the pH drops below 6.6 there is a significant inhibition of the methanogenic bacteria, and the acid conditions of a pH of 6.2 are toxic to these bacteria. At this pH, however, acid production will continue, since the acidogenic bacteria will produce acid until the pH drops to 4.5-5.0. Under balanced digestion conditions, the biochemical reactions tend to maintain the pH in the proper range automatically. Although the volatile organic acids produced during the first stage of the fermentation process tend to depress the pH, this effect is counteracted by the destruction of volatile acids and reformation of bicarbonate buffer during the second stage. If imbalance develops, however, the acid formers outpace the methane formers and volatile organic acids build up in the system. If imbalance continues, the buffer capacity may be overcome and a precipitous drop in pH may occur. As has been noted previously, buffering with lime may be necessary; other agents such as ammonium hydroxide may be used as buffering agents, but care must be exercised since excess ammonia, as well as the ammonium ion, can be toxic.

A number of other materials may be toxic to the methane bacteria, and this toxicity results in a reduction—sometimes to zero—of gas production. The stage 1 and 2 bacteria may be equally sensitive to certain toxins, but the response may not be as apparent. Usually the cessation of gas production is accompanied by an increase in organic acid accumulation; consequently the pH will drop, as discussed above. Common toxins include ammonia (> 1,500-3,000 mg/l of total ammonia nitrogen at pH > 7.4); ammonium ion (> 3,000 mg/l of total ammonia nitrogen at any pH); soluble sulfides (> 50-100 mg/l, possibly > 200 mg/l); and soluble salts of metals such as copper, zinc, and nickel. Many organic materials exhibit inhibitory effects.

Alkali and alkaline-earth metal salts, such as those of sodium, potassium, calcium, or magnesium, may be either stimulatory or inhibitory, depending on the concentration. These concentrations have been quantified, and the stimulatory and inhibitory ranges for these salts are shown in Table II-3. The toxicity is associated with the cation rather than the anion portion of the salt.

When combinations of these cations are present, the nature of their combined effect becomes quite complex. Some cations act as antagonists and reduce the toxic effects of other cations while others act synergistically, increasing the toxicity of other cations.[7]

It should be recognized that only the materials in solution are potential toxins. When a substance is removed from solution it cannot enter the bacterial cell and thus cannot affect the metabolism of an organism. For example, soluble sulfides and heavy metal ions are both toxic to biological activity; in combination, however, they combine to form very insoluble sulfide salts.

The magnitude of the toxic effect produced by a substance can often be reduced significantly if the concentration of the substance is increased gradually to allow the bacteria to acclimatize. Some tolerance to soluble sulfides and ammonia, for example, can be achieved through acclimatization.

Concentrations of soluble sulfide varying from 50 to 100 mg/l can be tolerated with little or no acclimatization required. With some acclimatization, concentrations up to 200 mg/l may be tolerated. Concentrations above 200 mg/l, however, are quite toxic. The soluble sulfide concentration in a digester is a function of the incoming sources of sulfur, the pH, the rate of gas production, and the availability of heavy metals to act as precipitants.

Ammonia is formed in the anaerobic process from the degradation of protein or urea. Ammonia is the source of nitrogen for the bacteria involved in the anaerobic process and at low concentrations it is stimulatory to the biological process. At the high concentrations in which it occurs in the digestion of some livestock and industrial wastes, however, its effects may be toxic.

The ammonium ion exists in equilibrium with dissolved ammonia gas ($NH_4^+ \rightleftharpoons NH_3 + H^+$) and the latter is inhibitory at a much lower concentration than the ammonium ion. The effects of ammonia on the digestion process are dependent upon its concentration and upon the pH. At a low pH, the equilibrium is shifted toward the ammonium ion so that inhibition is related to the ammonium ion concentration. Under these conditions, toxicity occurs at ammonia concentrations above 3,000 mg/l. At higher pH values equilibrium shifts toward dissolved ammonia, and if the ammonia concentration is between 1,500 and 3,000 mg/l the ammonia gas concentration may become inhibitory.

Heavy metal ions such as copper, zinc, and nickel have been reported to be toxic at concentrations as low as a few ppm. Generally, however, concentrations up to 100 mg/l (approximately 100 ppm) or more cause no harm because they are complexed by sulfides.

TABLE II-3 Stimulatory and Inhibitory Concentrations of Alkali and Alkaline-Earth Cations[6]

| Cation | Concentration in mg/l | | |
	Stimulatory	Moderately Inhibitory	Strongly Inhibitory
Sodium	100–200	3,500–5,500	8,000
Potassium	200–400	2,500–4,500	12,000
Calcium	100–200	2,500–4,500	8,000
Magnesium	75–150	1,000–1,500	3,000

There are a number of organic materials that may inhibit the digestion process. Some of these, such as the alcohols, are toxic at high concentrations, but when introduced continuously at low concentrations, they can be degraded as rapidly as they are added. Others may be treated successfully by precipitation from solution. For example sodium oleate, a fatty-acid base for soap, has been found to be toxic to anaerobic digestion in concentrations over 500 mg/l.[6] The inhibitor can be precipitated as the calcium oleate salt by addition of calcium chloride, and concentrations of sodium oleate as high as 2,000–3,000 mg/l can be tolerated.

For all of these materials, toxicity is readily apparent as a reduction in gas production.

Temperature

Process temperatures directly affect process conditions by controlling microbial growth rates. As the mesophilic process temperature is decreased below the optimum range of 33°-38°C, the net microbial growth rate decreases and the minimum solids-retention time for process stability increases. The same holds true for a shift in temperature in the thermophilic range.

The methane bacteria are very sensitive to sudden temperature changes, and for optimum process stability the temperature should be controlled very carefully within a narrow range of the selected operating temperature; it should at least be protected from sudden temperature changes. The reaction rate is greater in the thermophilic range than in the mesophilic, and as a result any upset in the operation (e.g., accumulation of volatile acids) will occur more rapidly, thus giving the operator less time to detect and correct the upset. However, for the same reason, the recovery rate is also much quicker.

Gas Composition

In the conversion of carbohydrates to carbon dioxide and methane, equal volumes of each gas are produced. However, not all of the carbon dioxide produced is released as gas since it is water-soluble. Carbon dioxide also reacts with hydroxyl ion to form bicarbonate. The concentration of bicarbonate will be affected by alkalinity, temperature, and the presence of other materials in the liquid phase. Conditions that favor bicarbonate production will increase the percentage of methane in the gas phase.

Hydroxyl ion is produced primarily during the deamination of biodegradable protein during stages 1 and 2, as a result of the reaction of ammonia with water to produce ammonium hydroxide. Therefore, the protein content of the substrate will significantly affect the quantity of carbon dioxide found as bicarbonate. The amount of carbon dioxide that is associated with the liquid stream as bicarbonate is reduced by increases in temperature and by decreases in pH. The retention time of the liquid in the reaction process will also affect the proportion of carbon dioxide found in the liquid phase; digestion at shorter retention times, for a given substrate, will produce a gas with higher methane content and a liquid phase with more carbon dioxide.

Other System Parameters

In addition to temperature and gas composition, the other system parameters that must be controlled to provide optimum conditions for the biological process of digestion are organic loading rates, influent solids concentration, concentration of toxic substances, and nutrient content. These are discussed in detail in Part III, Chapter 3, Operation and Maintenance.

References

1. Maki, L. R. 1954. Experiments on the microbiology of cellulose decomposition in a municipal sewage treatment plant. *Antonie Van Leeuwenhock Journal of Microbiology and Serology* (Netherlands) 20:185-200.
2. Heukelekian, H. 1927. Decomposition of cellulose in fresh sewage solids. *Industrial and Engineering Chemistry* 19:928-930.
3. McBee, R. H. 1948. The culture and physiology of a thermophilic cellulose fermenting bacterium. *Journal of Bacteriology* 56:653-663.
4. Stranks, D. W. 1956. Microbiological utilization of cellulose and wood. I. Laboratory fermentations of cellulose by rumen organisms. *Canadian Journal of Microbiology* 2:56-62.
5. Jeris, J. S., and McCarty, P. L. 1965. The biochemistry of methane fermentations using C^{14} tracers. *Journal of the Water Pollution Control Federation* 37:178-192.
6. McCarty, P. L. 1964. Anaerobic waste treatment fundamentals. 3. Toxic materials and their control. *Public Works* 95(11):91-94.
7. Kugleman, I. J., and McCarty, P. L. 1965. Cation toxicity and stimulation in anaerobic waste treatment. I. Sludge feed studies. *Journal of the Water Pollution Control Federation* 37:97-116.

Suggested Reading

1. An excellent state-of-the-art review of anaerobic fermentation processes for the conversion of cellulosic waste materials to chemicals and fuels is to be found in Compere, A. L., and Griffith, W. L. 1975. *Anaerobic Mechanisms for the Degradation of Cellulose.* ORNL-5056. Oak Ridge, Tennessee: Oak Ridge National Laboratory.
2. Bryant, M. P.; Wolin, E. A.; Wolin, M. J.; and Wolfe, R. S. 1967. *Methanobacillus omelienskii,* a symbiotic association of two species of bacteria. *Archiv fur Mikrobiologie* 59 20-31.
3. Bryant, M. P.; Tzeng, S. F.; Robinson, I. M.; and Jeyner, A. E., Jr. 1971. Nutrient requirements of methanogenic bacteria. In *Anaerobic Biological Treatment Processes,* Advances in Chemistry Series No. 105, Robert S. Gould, ed., pp. 23-40. New York: American Chemical Society.
4. Hungate, R. E. 1950. The anaerobic mesophilic cellulolytic bacteria. *Bacteriological Reviews* 14:1-49.
5. Lawrence, A. W., and McCarty, P. L. 1965. Sulfide prevention of heavy metal toxicity in digestion. *Journal of the Water Pollution Control Federation* 37:392-406.
6. Pfeffer, J. T. 1968. Increased loadings on digesters with recycle of digested solids. *Journal of the Water Pollution Control Federation* 40:1920-1933.
7. _____. 1974. *Reclamation of Energy from Organic Refuse.* EPA-670/2-74-016. Cincinnati, Ohio: U.S. Environmental Protection Agency, National Environmental Research Center.
8. _____. 1974. Temperature effects on anaerobic fermentation of domestic refuse. *Biotechnology and Bioengineering* 16:771-787.
9. _____, and Liebman, J. D. 1974. *Biological Conversion of Organic Refuse to Methane.* Report No. UILU-ENG-74-2019. Urbana, Illinois: University of Illinois, Department of Civil Engineering.
10. Reese, E. T.; Siu, R. G. H.; and Lenieson, H. S. 1950. The biological degradation of soluble cellulose derivatives and its relationship to the mechanism of cellulose hydrolysis. *Journal of Bacteriology* 59:485-497.

Chapter 2

Raw Materials and
Their Preparation

The raw materials that can be considered as substrates for the methane-generating bioconversion process are naturally occurring organic materials, generally cellulosic in nature. The biological mechanism by which the carbohydrates in these cellulosic materials are converted to methane is described in the previous chapter. Nearly all more-or-less natural* organic waste materials contain adequate amounts of the nutrients essential for the growth and metabolism of the anaerobic bacteria involved in this process. However, the chemical composition of these materials and the biological availability of the nutrients they contain vary not only with species and with factors affecting growth, but also with age.

Characteristics

The diverse nature of the potential raw materials for methane generation is described in Table I-1. These fibrous materials may be residues that result from other uses or they may be harvested directly as substrates for the bioconversion process. For example, crop residues may serve as a substrate; waste paper, being nothing more than wood fiber that has been subjected to various mechanical and chemical treatments, may also be used as a source of cellulose for this process. The substrate most widely used, however, is animal manure, which contains a substantial quantity of natural fibers that have been subjected to biochemical and mechanical treatments by the animal. These materials may receive little or no treatment prior to use in this process. Details on gas production from the materials suggested will be given in Part III.

*The phrase "more-or-less natural" is meant to include the residues and waste products of plant and animal life and such materials resulting from their processing as paper and food processing wastes; it does not include the organic wastes from fossil-fuel refining or the chemical process industry. It should be noted, however, that paper is lacking in essential nutrients such as N and P, and can serve as a source of cellulose only.

Materials of Plant Origin

1. Crop Residues

The variation in composition of plant materials is best illustrated by the examples listed in Table II-4, taken from Waksman (1952).[1] Waksman pointed out that the content of water-soluble substances—sugars, amino acids, proteins, mineral constituents—decreases with the age of the plant, while the content of cellulose, lignin, and (to a lesser extent) hemicelluloses and polyuronides (pectins, gums, and mucilages) increases. Although the actual nutrient content, with the possible exception of nitrogen, is seldom low enough to limit the rate of the anaerobic digestion process, the biological availability of the organic matter is nearly always a rate-limiting factor. Thus, although sugars are present as components of cellulose, until the cellulose is hydrolyzed the methane-generation process cannot proceed. Furthermore, the more lignin present to protect the cellulose from bacterial action, the less cellulose will be available for digestion. The efficiency of utilization of raw materials is therefore determined by the overall digestibility of the substrate, which is primarily a function of the lignin content.*

In the anaerobic decomposition of nonleguminous plants, the low availability of nitrogen is nearly always a rate-limiting factor.[2] The nitrogen content of crop plant residues varies with species, age of the plant, environmental factors affecting plant growth, and soil nitrogen availability. Experiments have shown, however, that only about 6 milligrams of nitrogen per gram of substrate (0.6 percent) is needed to sustain the anaerobic digestion process.[2] This means that nearly all stover, straw, chaff, or grain hulls contain enough nitrogen to eliminate its consideration as a rate-limiting factor in the anaerobic process. In spite of this, as will be pointed out in a later chapter, the C:N ratio in the raw material feed must be held within reasonable limits for the production of methane to proceed efficiently.

In rural areas of many developing countries, crop residues are used as animal fodder and so may not be available in sufficient quantity to generate enough methane to meet human needs. In areas where crop residues are not used for fodder, the quantity potentially available for methane generation can be estimated from known harvest weights by using certain crop residue coefficients. These coefficients, some of which are given in Table II-5, are dependent on the nature of the crop produced. The quantity of crop residue available for the digester is estimated by multiplying the known harvest weight by the residue coefficient. If a part of the crop residue is used as fodder, or for fiber, the net quantity of the crop residue actually available for methane generation depends, of course, on the amount remaining.

---

*For the purposes of the anaerobic fermentation process, lignin may be regarded as the fraction of the plant material that is nonbiodegradable; being insoluble in water, ether, alcohol, dilute alkali, and 70 percent (w/v) sulfuric acid, it remains insusceptible to bacterial attack.

TABLE II-4 Chemical Composition of Plant Material[a]

Constituent	Young Rye Plants	Mature Wheat Straw	Soybean Tops	Alfalfa Tops	Young Corn-stalks	More Mature Corn-stalks	Young Pine Needles	Old Pine Needles	Oak Leaves, Green	Oak Leaves, Mature, Brown
					Percent of Air-Dry Material					
Fats and waxes	2.35	1.10	3.80	10.41	3.42	5.94	7.65	23.92	7.75	4.01
Water-soluble constituents	29.54	5.57	22.09	17.24	28.27	14.14	13.02	7.29	22.02	15.32
Hemicelluloses	12.67	26.35	11.08	13.14	20.38	21.91	14.68	18.98	12.50	15.60
Cellulose	17.84	39.10	28.53	23.65	23.05	28.67	18.26	16.43	15.92	17.18
Lignin	10.61	21.60	13.84	8.95	9.68	9.46	27.63*	22.68	20.67	29.66
Protein	12.26	2.10	11.04	12.81	2.61	2.44	8.53	2.19	9.18	3.47
Ash	12.55	3.53	9.14	10.30	7.40	7.54	3.08	2.51	6.40	4.68

*The higher lignin content in the younger pine needles is due to the fact that this preparation has not been extracted with alcohol and is thus high in oils and waxes (which, in the procedure used, are measured as lignin).
[a]Source: Waksman (1952).[1]

TABLE II-5   Residue Coefficients for Major Crops[a]

Crop	Coefficient Range[b]
Soybean[c]	0.55 - 2.60[d]
Corn	0.55 - 1.20[e]
Cotton[c]	1.20 - 3.00[f]
Wheat[c]	0.47 - 1.75
Barley[c]	0.82 - 1.50
Paddy (unhusked rice, early varieties)	0.38 - 1.25[g]
Rye[c]	1.20 - 1.95
Oats[c]	0.95 - 1.75
Grain sorghum[c]	0.50 - 0.85
Sugar beet	0.07 - 0.20
Sugarcane[c]	0.13 - 0.25

[a]Information derived from the International Research and Technology Corporation[3] and Makhijani and Poole.[4]

[b]Residue coefficient is the ratio of the weight of *dry matter* of residue to recorded harvested weight at field moisture. For grains and straw, field moisture content is assumed to be 15 percent. In general, the lower coefficient stems from Reference 3 and the upper coefficient from Reference 4.

[c]Upper residue coefficients were determined by individual crop experts and supplied by Dr. Robert Yeck of the Agricultural Research Service of the United States Department of Agriculture, Washington, D.C., to Makhijani and Poole.[4] Small-grain estimates were for straw, and an additional factor of 0.25 for chaff has been included.

[d]The upper estimate is based on work by Alich and Hinman.[5] Estimates by the USDA [(c) above] are 0.85-2.6. Soybean is taken to be representative of other legumes.

[e]The upper estimate for corn residues is based on figures from Reference 5.

[f]Both lint and seed are included in the harvested crop.

[g]The upper estimate is based on data for India from a report by the Fertilizer Association of India.[6] When agricultural statistics are used to determine residues, it is important to note whether the data are for rice or paddy. In this case, rice hulls are included as residue—for early-maturing varieties they amount to about 30 percent of the total residue. (For late-maturing varieties the figure is about 20 percent.)

2. Paper Wastes

It is unlikely that waste paper will constitute a significant fraction of the raw-material feed for anaerobic digesters in rural areas of developing countries. Consequently, discussion of this source of cellulose will be limited to noting that the digestibility of papers is generally quite high, since the manufacturing process removes a significant portion of the lignin present in the initial wood. Thus, paper can serve as a source of carbon in the anaerobic digestion process. However, the digestibility of paper can be improved if the raw feed has a C:N ratio of 30:1. This can be achieved generally when paper is mixed with animal wastes in appropriate proportions. Therefore, if special circumstances make paper wastes available, they should be included in the raw-material feed.

3. Miscellaneous Plant Products

Several other plant products, although not normally considered as crop residues, are potential feed materials for methane generators. Forest litter, for example, is still abundant in the bush country in some parts of the world and, if circumstances justify its transportation, might well be included in the feed-stock.

In many parts of the tropical and semitropical world, the growth of filamentous algae, water hyacinth, and other aquatic weeds is a perennial problem in ponds, tanks, and other water bodies that have become eutrophic. These aquatic growths are good resources for energy production; for example, anaerobic digestion of 1 pound of algae (dry weight) can produce enough methane to yield about 6,000 Btu.[7]

Materials of Animal Origin

1. Manure

The animal wastes in an agricultural community represent a significant source of substrate for the bioconversion process. Indeed, as was pointed out in Part I, General Overview, manure is the major feedstock for most of the digesters in the developing world.

Besides being a source of carbon, these wastes are a potential source of the nitrogen required for the successful operation of an anaerobic fermentation process. The quantity, characteristics, and nitrogen and phosphorus contents of wastes (feces and urine) generated by livestock are shown in Table II-6.

The table clearly illustrates that the quantity and composition of animal wastes are dependent on the type of animal; poultry produce more volatile solids, nitrogen, and phosphorus, per unit weight of animal, than any of the other animals, for example. In addition, the effect of feed ration on the composition of manure is illustrated by comparing manure production from beef and dairy cattle. Thus the data in Table II-6, based on animals receiving rations typical of those used in the United States, are not necessarily typical of animals on significantly different feeds; an animal feeding on grass only will probably show much lower nitrogen content in the manure and urine. For these reasons, any design based on animal manure should be undertaken only after data on manure production, from the animals to be used, have been collected.

2. Night Soil

Human feces and urine—night soil—constitute another raw material that can be used as a substrate for production of methane by anaerobic fermenta-

TABLE II-6 Manure Production and Composition[a]

Animal	Daily Production				Composition		
	Per 1,000-lb live animal		Per 500-kg live animal		Volatile Solids[b]	Nitrogen	Phosphorus
	Volume (ft³)	Wet Weight (lb)	Volume (m³)	Wet Weight (kg)	Percent of Wet Weight		
Dairy cattle	1.33	76.9	0.038	38.5	7.98	0.38	0.10
Beef cattle	1.33	83.3	0.038	41.7	9.33	0.70	0.20
Swine	1.00	56.7	0.028	28.4	7.02	0.83	0.47
Sheep	0.70	40.0	0.020	20.0	21.5	1.00	0.30
Poultry	1.00	62.5	0.028	31.3	16.8	1.20	1.20
Horses	0.90	56.0	0.025	28.0	14.3	0.86	0.13

[a]Adapted from Fogg (1971).[8]

[b]See footnote, Part III, Chapter 1, p. 63.

tion. A typical composition is shown in Table II-7, from which it can be seen that night soil is quite similar in nitrogen and phosphorus content to animal manure. The public health benefits to be gained from anaerobic digestion of night soil and using the sludge as fertilizer, rather than spreading night soil directly on the land, are detailed in Part III, Chapter 6.

Collection, Preparation, and Storage

An important consideration in the generation of methane from agricultural and other wastes is the collection, preparation, and storage of the raw materials to be used in the anaerobic digestion process. In labor-intensive economies, methods should be considered that utilize available human and animal resources for the handling and processing of these wastes. Since the intent of anaerobic digestion is to produce an energy source, it is counterproductive to consider methods that require fossil fuel or other conventional sources of energy for handling and processing these wastes unless there is a significant net benefit.

The collection and processing of raw waste materials depend on their nature, and the quantities in which they are available may vary from country

TABLE II-7 Chemical Composition of Night Soil (Percent)[a]

Moisture	Organic Matter	Ash	Nitrogen	Phosphorus	Potassium	Sodium Chloride
95	3.4	1.6	0.57	0.052	0.22	1.02

[a]Adapted from Egawa (1975).[9]

to country and from region to region within a country. Hence the method for collecting and handling the waste materials may vary.

Because of the diverse nature of the raw materials that can be used for methane generation, they can be solid, liquid, or semisolid in nature; thus, appropriate methods are needed for their processing and subsequent utilization for methane production.

It has been pointed out that waste materials such as spent straw, hay, sugarcane trash, corn and corn-related plant stubble, and bagasse can be used for the generation of methane. However, in order to obtain more gas from them and to facilitate their flow through the digester and its appurtenances, it is advisable to shred them into small pieces. In rural areas, manually operated shredders should be used because they are less susceptible to failure and labor is available. These materials can be handled easily because they are relatively dry, but measures should be taken to reduce the fire hazard stemming from their use.

Forest litter is available abundantly in the bush country of some parts of the world. This material can be hauled in bags or on animal-drawn vehicles and processed along with other agricultural and human wastes at the site of the gas plant.

Where the production of large quantities of methane is possible, due to the availability of large amounts of human, animal, and agricultural wastes (as in the case of small farms and village dairy cooperatives), the digester (biogas plant) may be designed to receive wastes directly. In such situations it is common to use equilibrating/mixing chambers ahead of the digester, to maintain control of the composition of the feedstock. In situations where the raw material cannot be fed directly to the digester after collection, it may be stored briefly; mixed animal and agricultural wastes may be stacked in a storage shed prepared for this purpose. It should be covered and not used for prolonged storage to avoid rotting and fly breeding in the manure pile. At the most, quantities that are sufficient for a 2-day loading of the digester should be stored in this way.

The quality of animal manure can be substantially altered by exposure to the environment. In particular, rain will leach a significant portion of the soluble material from the manure and substantial losses in total nitrogen, for example—ranging from about 30 percent to as much as 85 percent of the excreted nitrogen—can be incurred.[10]

In some parts of the developing world, a great deal of flotsam and seaweed material is available in coastal village areas where the economy is primarily dependent on fishing. These are good raw materials for the production of methane. If there is enough inducement, the fishermen and other rural workers may be persuaded to collect the aquatic vegetation for charging the biogas plants along with other rural wastes.

References

1. Waksman, S. A. 1952. *Soil Microbiology*. New York. John Wiley and Sons.
2. Buswell, A. M., and Hatfield, W. D. 1939. *Anaerobic Fermentation*. Bulletin No. 32. Urbana, Illinois: Illinois State Water Survey.
3. International Research and Technology Corp. 1973. *Problems and Opportunities in Management of Combustible Solid Wastes*. Report No. EPA-670/2-73-056. Washington, D.C.: U.S. Environmental Protection Agency.
4. Makhijani, Arjun, and Poole, Alan D. 1975. *Energy and Agriculture in the Third World*. Report to the Energy Policy Project of the Ford Foundation. Cambridge, Massachusetts: Ballinger Press.
5. Alich, J. A., Jr., and Hinman, Robert E. 1974. *Effective Utilization of Solar Energy to Produce Clean Fuel*. Final report under Grant GI 38723, National Science Foundation. Menlo Park, California: Stanford Research Institute.
6. Saolpurkar, V. K., and Balkundi, S. V. 1969. *Rice*. New Delhi: Fertilizer Association of India.
7. Oswald, W. J., and Golueke, C. G. 1964. Solar power via a botanical process. *Mechanical Engineering* 80(2):40.
8. Fogg, C. E. 1971. Livestock waste management and the conservation plan, pp. 34-35. In *Livestock Waste Management and Pollution Abatement. The Proceedings of the International Symposium on Livestock Wastes, April 19-22, 1971, The Ohio State University, Columbus, Ohio*. St. Joseph, Michigan: American Society of Agricultural Engineers.
9. Egawa, Tomoji. 1975. Utilization of organic materials as fertilizer in Japan, pp. 253-271. In *Organic Materials as Fertilizers: Report of the FAO/SIDA Expert Consultation, Rome, 2-6 December 1974*. Rome: Food and Agriculture Organization of the United Nations.
10. Gilbertson, C. B.; McCalla, T. M.; Ellis, J. R.; and Woods, W. R. 1971. Characteristics of manure accumulation removed from outdoor, unpaved, beef cattle feedlots, pp. 56-59. In *Livestock Waste Management and Pollution Abatement: The Proceedings of the International Symposium on Livestock Wastes, April 19-22, 1971, The Ohio State University, Columbus, Ohio*. St. Joseph, Michigan: American Society of Agricultural Engineers.

Chapter 3

Product Storage and Use

Pure methane is a colorless and odorless gas. It generally constitutes between 50 percent and 70 percent of the gas produced by anaerobic digestion. The other 30-50 percent is primarily carbon dioxide, with a small amount of hydrogen sulfide. Table II-8 lists some of the important physical and chemical properties of methane.

Digester gas (biogas) burns with a blue flame and has a heat value ranging from about 500 to 700 Btu/ft^3 (22,000 to 26,000 kJ/m^3) when its methane content ranges from 60 to 70 percent. It can be used directly in gas-burning appliances for heating, cooking, lighting, and refrigeration or as a fuel for internal-combustion engines having a compression ratio of 8:1 or greater.

TABLE II-8   Selected Physical and Chemical Properties of Methane[a]

Chemical formula:	CH_4
Molecular weight:	16.042
Boiling point at 14.696 psia (760 mm)	$-258.68°F$ ($-161.49°C$)
Freezing point at 14.696 psia (760 mm)	$-296.46°F$ ($-182.48°C$)
Critical pressure:	673.1 psia (47.363 kg/cm^2)
Critical temperature:	$-116.5°F$ ($-82.5°C$)
Specific gravity:	
Liquid (at $-263.2°F$ [$-164°C$])	0.415
Gas (at 77°F [25°C] and	
14.696 psia [760 mm])	0.000658
Specific volume at 60°F (15.5°C) &	
14.696 psia (760 mm):	23.61 ft^3/lb (1.47 l/gm)
Calorific value 60°F (15.5°C) &	
14.696 psia (760 mm):	1,012 Btu/ft^3 (38,130.71 kJ/m^3)
Air required for combustion ft^3/ft^3:	9.53
Flammability limits:	5 to 15 percent by volume
Octane rating:	130
Ignition temperature:	1,202°F (650°C)
Combustion equation:	$CH_4 + 2O_2 \rightarrow CO_2 + 2H_2O$
O_2/CH_4 for complete combustion:	3.98 by weight
O_2/CH_4 for complete combustion:	2.0 by volume
CO_2/CH_4 from complete combustion:	2.74 by weight
CO_2/CH_4 from complete combustion:	1.00 by volume

[a]Sources: Katz, et al. (1959);[1] Johnson and Auth (1951);[2] and Weast, et al. (1964).[3]

44

The gas is most easily used in burning appliances that can use it directly from low-pressure collection units. In this case, a small digester with a floating cover can provide limited gas storage. Alternatively, gas from a digester with a fixed cover can be piped for collection in an auxiliary gas holder with a floating cover. A gas delivery line is connected to the gas storage unit with an on-off control valve. This line must include a flame trap of the P-type, used in domestic plumbing lines, between the gas storage unit and the appliance being used. Systems of this type are in use in India, Taiwan, and other warm areas.

Consumption rates for gas-burning appliances, as given by Singh,[4] are listed in Table II-9.

Digester gas may be used as fuel for motor vehicles, but its performance efficiency is dependent on the methane content—that is, the content of carbon dioxide (CO_2) and hydrogen sulfide (H_2S). Furthermore, if enough fuel is to be carried in the vehicle to permit reasonable distances of travel, the gas must be "concentrated" by being compressed and stored in high-pressure cylinders. (For local travel, low-pressure bag storage can be used.) Thus, the

TABLE II-9   Quantities of Biogas Required for a Specific Application[a]

| Use | Specification | Quantity of Gas Required | | Reference No. |
		ft^3/hr	m^3/hr	
Cooking	2″ burner	11.5	0.33	5
	4″ burner	16.5	0.47	5
	6″ burner	22.5	0.64	5
	2″–4″ burner	8-16	0.23-0.45	6
	per person/day	12-15+	0.34-0.42+	6
	per person/day	12+	0.34+	7
Gas lighting	per lamp of 100 candle power	4.5	0.13	7
	per mantle	2.5	0.07	6
	per mantle	2.5-3.0	0.07-0.08	5
	2 mantle lamp	5	0.14	5
	3 mantle lamp	6	0.17	5
Gasoline or diesel engine[b]	converted to biogas, per hp	16-18	0.45-0.51	6
Refrigerator	per ft^3 capacity	1	0.028	5
	per ft^3 capacity	1.2	0.034	6
Incubator	per ft^3 capacity	0.45-0.6	0.013-0.017	5
	per ft^3 capacity	0.5-0.7	0.014-0.020	6
Gasoline	1 liter	47-66[c]	1.33-1.87[c]	6
Diesel fuel	1 liter	53-73[c]	1.50-2.07[c]	6
Boiling water	1 liter	3.9[d]	0.11[d]	5

[a]Adapted from Singh (1972).[4]
[b]Based on 25 percent efficiency.
[c]Absolute volume of biogas needed to provide energy equivalent of 1 liter of fuel.
[d]Absolute volume of biogas needed to boil off 1 liter of water.

use of digester gas in internal-combustion engines requires special equipment and processes, which further include:

- Reducing the H_2S content of the gas to less than 0.25 percent to prevent corrosive damage to metal surfaces, particularly bearings and other working parts.[8]
- A scrubbing system to remove CO_2. While carbon dioxide exerts no harmful effects on internal-combustion engines, the presence of this non-combustible gas (raw digester gas contains approximately 30 percent CO_2 by volume) reduces the heat content per unit of gas volume and thus lowers the operating efficiency of the engine.
- A compressor capable of compressing gas to pressures between 2,000 and 3,000 psi (140 and 210 kg/cm^2). When digester gas is to be compressed, it is imperative that the carbon dioxide be removed to prevent mechanical damage to the compressor caused by liquefaction of the CO_2. (See Table II-10 for critical constants of digester gas constituents.)
- High-pressure cylinders, of the type used for oxygen, that can safely store gas at pressures up to 2,400 psi (170 kg/cm^2), to be installed on the vehicle.
- A set of similar high-pressure storage cylinders equipped with a pressure control panel for filling the vehicle cylinders.
- Pressure-reducing valves installed on the vehicle to supply the low-pressure gas required by the carburetor. Usually two valves are used, one a high-pressure reduction valve to reduce the tank pressure (2,400 psi, 170 kg/cm^2) to an intermediate low pressure of, say, 50 psi (3.5 kg/cm^2), and a second to reduce the pressure below atmospheric pressure so that gas will not escape from the line when the engine is not operating.
- An automatic gas/air mixing valve installed on the air intake of the carburetor.

TABLE II-10   Critical Temperatures and Pressures of Digester Gas Constituents[a]

Gas	Critical Temperature		Critical Pressure	
	°F	°C	Psia	kg/cm^2
CH_4	−115.8	−82.1	673	47.3
CO_2	87.8	31.0	1,072	75.3
H_2S	212.7	100.4	1,307	91.9
NH_3	270.5	132.5	1,654	116.3

[a] Adapted from data in: Weast, Robert C., ed. 1975. *Handbook of Chemistry and Physics.* 56th ed. Cleveland, Ohio: CRC Press. The critical temperature is the temperature above which a gas cannot be liquified by pressure. The critical pressure is the pressure at which a gas is in equilibrium with its liquid at the critical temperature.

Conversion kits are commercially available for installation on standard internal-combustion engines to permit their operation on either gasoline or methane. One such conversion kit, installed on a half-ton pickup truck powered by a six-cylinder engine at the University of Manitoba in Canada, gave a fuel-consumption value of 11 ft^3 of methane per mile (19.3 l/km).[9,10,11]

Use of raw digester gas or scrubbed methane in internal-combustion engines can be considered only where large digesters have been installed to produce large volumes of gas. It must be determined whether the cost of the equipment needed for cleaning, compressing, and using the gas can be justified when compared with the cost of standard petroleum fuels. Operation of stationary engines should have economic priority over mobile vehicles because there is no need for compressing equipment and high-pressure storage tanks. Scrubbing equipment to reduce hydrogen sulfide to a tolerable level may be necessary, however, to eliminate corrosion of bearings.

References

1. Katz, D. L.; Cornell, D.; Kobayashi, R.; Poettmann, J. A. Vary; Elenbass, J. R.; and Weinaug, C. F. 1959. *Handbook of Natural Gas Engineering*. New York: McGraw-Hill.
2. Johnson, A. J., and Auth, G. H. 1951. *Fuels and Combustion Handbook*. New York: McGraw-Hill.
3. Weast, Robert C.; Selby, Samuel M.; and Hodgman, Charles D., eds. 1964. *Handbook of Chemistry and Physics*. 45th ed. Cleveland, Ohio: The Chemical Rubber Co.
4. Singh, Ram Bux. 1972. The bio-gas plant: generating methane from organic wastes. *Compost Science* 13(1):20-25.
5. _____. 1971. *Bio-Gas Plant: Generating Methane from Organic Wastes*, p. 36. Ajitmal, Etawah (U.P.), India: Gobar Gas Research Station.
6. Fry, L. J., and Merrill, R. 1973. *Methane Digesters for Fuel, Gas and Fertilizer* Newsletter No. 3. Santa Cruz, California: New Alchemy Institute–West.
7. Khadi and Village Industries Commission. *Gobar Gas–Why and How*. n.d. Bombay, India: Khadi and Village Industries Commission.
8. Abell, C. 1952. *Butane-Propane Power Manual: Principles of LP-Gas Carburetion*. Los Angeles, California: Jenkins Publications, Inc.
9. Lapp, H. M.; Schulte, D. D.; Sparling, A. B.; and Buchanan, L. C. 1975. Methane production from animal wastes. 1. Fundamental considerations. *Canadian Agricultural Engineering* 17(2):97-102.
10. See also: Wong, J. K. S. 1976. *Studies of Mixtures of Methane and Carbon Dioxide as Fuels in a Single Cylinder Engine [CLR]*. Mechanical Engineering Report No. MP-70. Ottawa: National Research Council of Canada.
11. Conversion kits are available from, for example, the following:
    IMPCO
    Division of A. J. Industries, Inc.
    16916 Gridley Place
    Cerritos, California 90701 USA

    Dual Fuel Systems, Inc.
    720 W. Eighth Street
    Los Angeles, California 90017 USA

Chapter 4

Residue: Composition, Storage, and Use

Much information has been collected and many reports written on methane production and the reduction and stabilization of volatile solids* from various kinds of animal, crop-residue, and food-processing wastes by anaerobic digestion.[1,2,3] However, most people working in this field have been concerned mainly with collecting information useful in the design and operation of anaerobic digesters, and have given very little attention to the characterization of the remaining residues (sludges). More specifically, sludges have not been characterized in terms useful in predicting their value as fertilizers or soil-amendment materials. For the most part, investigators have simply stated that the sludge produced by anaerobic digestion may have a fertilizer value greater than that of the original raw waste, that no offensive odors result when it is stored in lagoons or spread on land, and that rodents and flies are not attracted by the remaining solid or liquid residues.

The composition of the sludge produced by anaerobic digestion is determined by the composition of the raw material fed to the digester. (See Part II, Chapter 2.) As a working premise, it is commonly assumed that 70 percent of the organic constituents available for decomposition under conditions favorable for anaerobic digestion is decomposed. On this basis, the organic fraction of the sludge produced from the digestion of plant and animal waste may be expected to contain approximately 30 percent of the original weight of the organic material added to the digester. This organic fraction of anaerobically produced sludges consists of three classes of material: original substances protected from bacterial decomposition by lignin and cutin, newly synthesized bacterial cellular substances, and relatively small amounts of volatile fatty acids. The amount of bacterial cell mass is small because anaerobic cultures typically convert only about 10-20 percent of the carbon substrate to cells. Cultures adapted to aerobic systems, on the other hand,

---

*Volatile solids are that portion of the total solids that is volatilized by combustion, usually under certain standard conditions. See footnote, p. 63.

generally convert about 50 percent of the carbon substrate to cell mass. Thus, waste materials are stabilized by removal of most of the carbon that would otherwise be used for bacterial growth, leaving few bacterial cells. This is a major reason that there is less risk of odor or insect breeding when sludges produced by anaerobic digestion are stored and spread on land than there is with aerobically treated organic waste materials. This is also the main reason that nutrient deficiency of mature plant residues—despite their decreased content of sugars, amino acids, proteins, and mineral constituents and increased content of cellulose, lignins, and hemicellulose—is seldom great enough to limit the growth of the anaerobic bacterial populations and, by so doing, become the rate-limiting factor in the anaerobic process.

Anaerobic digestion of plant residues and animals wastes conserves nutrients needed for the continued production of crops; the only materials removed from the system for purposes other than putting back on the land are the gases generated—CH_4, CO_2, and H_2S. The fate of nitrogen is particularly important; a major advantage of anaerobic digestion of plant residues and animal wastes is the conservation, in organic or ammonium nitrogen forms, of practically all the nitrogen present in the materials used. This is demonstrated by the example outlined in Table II-11, which shows the results of experiments with an anaerobic system using dung as the feed material.[4]

These results show that before the spent slurry is treated, approximately 99 percent of the nitrogen* in the original material has been retained; about 1 percent or less is lost in the gases evolved in the process.

The distribution of total nitrogen in the sludge, between organic and ammonium nitrogen, depends on the distribution in the raw material. In the example in Table II-11, more than 17 percent of the total original nitrogen is in the form of ammonia after the digestion, with a small amount (approximately 0.8 percent) in the form of other volatile compounds. In another example, anaerobic digestion of rice straw, with its lower nitrogen content (about 0.8 percent[5]), produced only 8-10 percent in the form of ammonia nitrogen.[6] Thus, as much as 18 percent of the original nitrogen can be lost by handling the slurry in a way that allows loss by volatilization. Since only a very small amount of the nitrogen contained in plant residues and animal waste will be utilized by anaerobic bacteria to synthesize protein, the greater the nitrogen content of the material introduced into anaerobic digesters, the greater will be the concentration of ammonium nitrogen in the sludge. Where nitrogen economy is important, it follows that the higher the nitrogen content of feed materials used in anaerobic digestion, the greater the need for proper storage and proper application of sludges to the land to reduce nitrogen loss by volatilization. This will ensure that nearly all the nitrogen present

---

*Throughout this chapter, references to nitrogen content are based on the Kjeldahl analysis; thus, what is meant is total Kjeldahl nitrogen, or "TKN."

TABLE II-11   Nitrogen Balance of Anaerobic System[a]

	Total Kjeldahl Nitrogen (TKN) (g)	Ammonia Nitrogen (g)	Organic Nitrogen (g)
Nitrogen input:			
In 1.5 kg dung (wet)	3.805	0.130	3.675
In 500 ml inoculum	0.798	0.126	0.672
Total input	4.603	0.256	4.347
(percent)	(0.31)	(0.017)	(0.29)
Nitrogen recovered:			
In gases evolved	0.005	0.005	–
In digested slurry	4.536	0.797	3.739
Total recovered	4.541	0.802	3.739
(percent of total TKN input)	(98.7)	(17.4)	(81.2)
Nitrogen recovered:			
In digested slurry after			
evaporation to dryness	3.731	0.028	3.703
(percent of total TKN input)	(81.1)	(0.006)	(80.4)

[a]Adapted from Idnani and Varadarajan (1974).[4]

in plant residues will sooner or later be available for use by succeeding crops. To minimize ammonia nitrogen losses, digested sludge should be stored in deep lagoons or tanks that present a minimum of surface area for ammonia volatilization. Nitrogen is conserved to the greatest extent if the anaerobically digested sludge is injected below the soil surface a few days before crop planting or just prior to cultivation. If the sludge is spread on the soil surface and allowed to dry without interruption by rainfall, nearly all the ammonia nitrogen will be lost by volatilization.

One of the benefits of harvesting plant residues for the generation of methane is the saving of valuable plant nutrients that might otherwise be lost by leaching to depths below the root zone and beyond availability for absorption by crop plants. Many nutrient elements are precipitated in soils to form sparingly soluble compounds, and thus, with time, they become less available to plants. Some elements, such as potassium, are easily leached from plant residues left to decompose on the soil surface. On permeable soils, these elements may be leached below the root zone of the next or the succeeding crop. Other nutrients, such as phosphorus and some trace-metal elements, mineralized from plant residues during decomposition at or in the soil surface, often become less available for plant absorption with time as they react with inorganic soil constituents to form compounds with relatively low solubilities. Therefore, if plant residues are harvested and properly stored as feedstock for an anaerobic digester, many essential plant nutrients are used more efficiently because they will be returned to the soil just before planting of the next crop. Many of the trace elements that are chelated during the microbial decomposition of plant residues in the digesters will be maintained in this highly available state until they can be used by the succeeding crop. If the exposure period of these elements to the soil environment is shortened by their storage as constituents of sludge, plant nutrition will be much improved.

In many instances, entomologists and pathologists recommend the plowing under of plant residues to decrease the incidence of insect and plant pathogen infestation in the succeeding crop. On the other hand, where soils are subject to severe erosion by water or wind, soil conservationists recommend that plowing be delayed as long as possible before planting, or be eliminated altogether. Harvesting and digesting plant residues may be one way of reconciling differences in recommendations given by specialists. Anaerobic digestion of plant residues may be a way of controlling some insects and diseases more effectively than would be possible in residues decomposed in aerobic soil environments. For example, some scientists believe fungi that cause plant diseases could be more effectively controlled by anaerobically digesting plant residues than by plowing the residues under the soil.

Since the aerobic surfaces of cultivated soils contain abundant supplies of various kinds of microorganisms, readily degradable organic constituents of digested sludge are rapidly decomposed to CO_2 and H_2O or are transformed to humus compounds. The end result of applying anaerobically digested sludge on soils is the same as that occurring with any other kind of compost; the humus materials formed improve soil properties such as aeration and moisture-holding capacity and increase cation-exchange capacity, water-infiltration capacity, etc. Furthermore, the sludge serves as a source of energy and nutrients for the development of microorganism populations that directly and indirectly favor the solubility—and thus the availability to higher plants—of essential nutrients contained in soil minerals.

Even though the organic fraction of anaerobically digested sludges contains a relatively large proportion of lignin and lipid substances, there will be some degradation of this fraction in the soil. In a study where anaerobically digested municipal sludge has been applied annually for 6-7 years at various rates, on different soil types, the soil organic matter reached a more-or-less dynamic equilibrium state.[7] It appears that about 20-40 percent of the organic matter in the sludge decomposes each year, depending on difference in soil type. Within 5 years after sludge applications began, soil organic content reached a more-or-less stable level; maximum soil organic matter content at equilibrium depended on the amount of sludge applied annually. That is, the greater the amount applied each year, the greater the soil organic matter content when equilibrium was attained.

References

1. Buswell, A. M., and Hatfield, W. D. 1939. *Anaerobic Fermentation.* Bulletin No. 32. Urbana, Illinois: Illinois State Water Survey.
2. Smith, R. J. 1973. The Anaerobic Digestion of Livestock Wastes and the Prospects for Methane Production. Paper presented at the Midwest Livestock Waste Management Conference, 27-28 November 1973, Iowa State University, Ames, Iowa.
3. Lapp, H. M.; Schulte, D. D.; and Buchanan, L. C. 1974. *Methane Gas Production from Animal Wastes.* Publication 1528. Ottawa, Canada: Canada Department of Agriculture.
4. Idnani, M. A., and Varadarajan, S. 1974. *Fuel Gas and Manure by Anaerobic Fermentation of Organic Materials.* Technical Bulletin 46. Rev. ed. New Delhi: Indian Council of Agricultural Research.
5. Calliban, C. D., and Dunlap, C. E. 1971. *Construction of a Chemical-microbial Pilot Plant for Production of Single-cell Protein from Cellulosic Wastes*, pp. 15-21. Publication No. SW 24C. Washington, D.C.: U.S. Environmental Protection Agency.
6. Acharya, C. N. 1935. Studies on the anaerobic decomposition of plant materials. I. The anaerobic decomposition of rice straw (*Oryza sativa*). *Biochemical Journal,* Part I, 29:528-540.
7. For unpublished results (1975) contact T. D. Hinesly and R. L. Jones, Department of Agronomy, University of Illinois, Urbana, Illinois.

Chapter 5

Public Health Aspects

The public health aspects of the production of methane by the anaerobic digestion of wastes, under conditions to be encountered in a developing country, can be adequately discussed in terms of potential hazards and precautionary measures.

Potential Hazards

The potential hazards inherent in the anaerobic digestion of wastes are the result of two practices—the handling involved in the use of human feces (night soil) as a part of the waste fed to the digesters, and the use, in crop production, of the sludges produced in these digesters as fertilizers. Although the use of wastes from diseased animals may entail some danger (e.g., leptospirosis),[1,2] it would be less than that involved in the use of human feces.

The nature and variety of diseases that can be transmitted through improper handling of human excrement, together with their causative agents, are well documented in textbooks on clinical bacteriology and parasitology.[3,4] These diseases can be of viral, bacterial, protozoan, and helminthic origin; some examples, together with their causative organisms, are given in Table II-12.

The public health hazards associated with the use, as fertilizer, of sludge from an anaerobic digester, when untreated or minimally treated human excreta constitute part of the raw material feed, depend on these factors:

- The incidence of viable pathogenic organisms found in fecal waste material;
- The survival rate of these organisms in the sludge; and
- The storage time of the sludge prior to its application to the land.

These health hazards may be assessed on the basis of available information on the occurrence and survival of pathogenic organisms in raw sewage digesters and in sludge used in the field.

TABLE II-12   Some Diseases and Causative Organisms of Concern in Handling of Human Excrement.

Category	Disease	Organisms (where identified)
Viral	Infectious hepatitis	
	Gastroenteritis	
	Respiratory illness	adenovirus
		reovirus
	Poliomyelitis	enterovirus (poliovirus)
Bacterial	Typhoid fever	*Salmonella typhosa*
	Salmonellosis	*Salmonella* spp.
		(Ex.. *S. paratyphi,*
		S. schottmuelleri)
	Bacillary dysentery	*Shigella* spp.
	(Shigellosis)	
	Cholera	*Vibrio cholerae*
	Tuberculosis	*Mycobacterium tuberculosis*
Protozoan	Amebiasis	*Entamoeba histolytica*
	(Amebic dysentery)	
Helminthic	(Roundworm)	*Ascaris lumbricoides*
	(Pinworm)	*Oxyaris vermicularis*
	(Whipworm)	*Trichurus trichiura*
	(Tapeworm)	*Taenia saginate*
	(Hookworm)	*Ancylostoma duodenale*
		Necator americanus

Survival in the Digester

The published data on survival of pathogenic microorganisms in the anaerobic digestion process have a wide range of values. Survival data for some of the more important enteric microorganisms listed in Table II-12 are given in Table II-13. These examples, together with other data available, demonstrate the importance of the anaerobic digestion process in the treatment of human wastes; with few exceptions, pathogenic enteric microorganisms are effectively killed off if the digestion time is not significantly shorter than 14 days at a temperature not significantly lower than 35°C.

Studies of viruses typical of the enteric viruses hazardous to man show that anaerobic digestion at 35°C for 14 days will result in a 99.9-percent die-off. Poliovirus type 1 (enterovirus), for example, has been found to suffer an inactivation rate of 90 percent per day (range 83.6-93.9 percent) when subjected to anaerobic fermentation at 35°C.[5]

Information on the survival of protozoa in anaerobic digestion processes is meager. In one study, *E. histolytica* cysts were reported not to survive anaerobic digestion.[9] However, in another study, a species of amoeba similar to *E. histolytica*, but having different cultural requirements, was found in sludge

TABLE II-13   Die-off of Enteric Microorganisms of Public Health Significance During Anaerobic Digestion

Organisms	Temperature (°C)	Residence Time (days)[a]	Die-off (%)	Reference
Poliovirus	35	2	98.5	5
Salmonella ssp.	22-37	6-20	82-96	6
Salmonella typhosa	22-37	6	99	6
Mycobacterium tuberculosis	30	Not reported	100	7
Ascaris	29	15	90	6
Parasite cysts	30	10	100[b]	8

[a]Time indicated is time of digestion.
[b]Does not include *Ascaris*.

digestion and Imhoff tanks.[10] The major exception to the pathogenic microorganisms effectively killed during anaerobic digestion is the roundworm, *Ascaris lumbricoides*. While other encysted helminths are completely destroyed, ascaris cysts are able to survive even after 14 days of anaerobic fermentation at 35°C.[6] Another study reported that thermophilic digestion destroyed the ascaris eggs completely. However, sludge from mesophilic digestion required additional storage and drying for about 6 months to destroy the ascaris eggs completely.[11] Nevertheless, since the die-off rate is expected to be sufficiently high (90 percent), anaerobic digestion of organic material for biogas production provides a public health benefit beyond that of any other treatment likely to be in use in rural areas of developing countries.

The gastrointestinal disorders caused by hookworm (*Ankylostoma*) are another public health concern endemic to many semitropical and tropical developing countries. It is likely that hookworm eggs may find their way into the digesters if they are fed with night soil. Studies of their survival in septic tanks constructed in West Bengal (India) suggested that viable hookworm eggs were recovered in the septic tank effluents, although high removal rates were achieved (90 percent).[12] Since biogas plants will similarly collect and process wastes, the risk of indiscriminate dissemination of eggs of parasitic organisms is minimized. Storage and drying after the digestion period will further minimize—and perhaps eliminate—the risk of parasite eggs spreading in the rural environment of developing countries.

Survival in the Field

"Effective" destruction of pathogenic enteric microorganisms does not preclude the survival of at least some microorganisms of public health significance—despite destruction of as much as 99+ percent of such organisms originally present. Once the sludge has been removed from the anaerobic system, however, those organisms that have survived the process continue to die off because of the lack of nutrients and the hostile environment. This die-off continues both during storage of the sludge and after the sludge has been applied to the soil. The possibility of contaminating crops with these surviving pathogens could be eliminated by pasteurization, which is the practice in some European countries; this process is so costly, however, that its economic feasibility is marginal, even for a developed country.

The survival of enteric microorganisms has been the subject of many investigations and the results, which depend on factors such as the organism, the ambient temperature, the type of soil, the moisture, and pH, have varied widely, ranging from a few hours (in soil) to several months.[13]

The hazard in the use of sewage sludge comes from contact of the sludge with the edible portion of the crop—the organisms in the soil cannot, of course, pass the cell-wall barrier of the plant root system. Because pathogens are removed from run-off or leachate that has percolated through rather short columns of soil (in some cases as little as 3 feet of soil is needed),[14] the hazard does not normally extend to groundwater drinking supplies that are protected from surface run-off. The magnitude of the public health hazard posed by the use of digester sludge as fertilizer depends on the concentration (number) of pathogenic organisms that survive digestion, and the die-off rate of these survivors during sludge storage and in the soil. Since the number that survive is proportional to the number originally present, the heavier the contamination in the raw material, the more serious the concern with the sludge. Nevertheless, *there is no other practical method of treating human excreta —whether for disposal or to return nutrients to the land as fertilizer—that will reduce the burden of pathogenic organisms as much as anaerobic digestion.*

In spite of the survival of some pathogens and parasites, the literature documents no disease outbreaks associated with the use of digested night soil and animal wastes in crop production.

The inhabitants of most villages or rural areas in developing countries are probably already exposed to the enteric diseases endemic to their area. The introduction of anaerobic digesters for night soil or animal wastes, therefore, should not create any new or additional health hazards; on the contrary, it should reduce the present health hazards significantly.

Precautionary Measures

A list of procedures could be drawn up that could effectively eliminate the dangers inherent in handling human wastes. However, many such directives would not be practical or economically feasible in developing countries; thus, it is difficult to conjure up a list of foolproof handling procedures. For example, a seemingly feasible precaution for workers in a village-scale methane production plant would be for them to scrub their hands with soap and water after each handling of night soil and to reserve one set of clothing solely for work use. Unfortunately, extra sets of clothing may not be practical at the village level. The situation is further aggravated by the fact that mechanized handling will be minimal, if used at all; this means increased human contact.

Perhaps precautionary measures could most easily be imposed on methods of collecting night soil, loading and unloading the digester, and using the residue. The most desirable situation would be to establish a system of village latrines that are directly connected to the digester. In this way, handling of night soil would be eliminated. Failing this, however, vessels used for transporting the wastes should be used exclusively for that purpose. Spillage should be avoided during transport. Storage could be minimized by operating the digester as a continuous culture, to the extent of loading it once each day. In all of the steps, the handler should avoid direct contact with any fecal material.

On the basis of current knowledge, it seems clear that using the sludge from unheated digesters as fertilizer will pose a much smaller health hazard than the present use of untreated night soil, because of the reduction in the number of pathogenic organisms in the anaerobic process. The use of a heated digester, however, will reduce the hazard considerably more, though not necessarily to zero.

In summary, without safeguards that would not be economically feasible in developing countries, some degree of health hazard, however minimal, would be involved in biogas production in such countries, at least insofar as human feces are used. However, the degree of hazard would be significantly less than that to which the people are currently exposed in the traditional disposal or use of night soil. Indeed, the institution of a biogas scheme could very well serve as a spur to the construction and use of household latrines, based on the economic value of the fuel (and fertilizer) obtained. In this way, the public health hazard of the common practice in rural areas of defecating in the fields would be minimized. Furthermore, connecting the latrines directly to household (or institutional) biogas generators would eliminate any health hazard of direct handling of human feces.[7]

References

1. McCalla, T. M., and Elliott, L. F. 1971. The role of micro-organisms in the management of animal wastes on beef cattle feedlots, pp. 132-134. In *Livestock Waste Management and Pollution Abatement: The Proceedings of the International Symposium on Livestock Wastes, April 19-22, The Ohio State University, Columbus, Ohio*. St. Joseph, Michigan: American Society of Agricultural Engineers.
2. Diesch, S. L.; Pomery, B. S.; and Allred, E. R. 1971. *Survival of Pathogens in Animal Manure Disposal*. Final Report, Grant EP-00302 EPA, University of Minnesota. Washington, D.C.: U.S. Environmental Protection Agency.

3. Noble, E., and Noble, G. 1971. *Parasitology*. 3rd ed. Philadelphia: Lea and Febiger.
4. Craig, C. F., and Faust, E. C. 1970. *Clinical Parasitology* Philadelphia: Lea and Febiger.
5. Bertucci, J.; Lue-Hing, C.; Zenz, D.; and Sedita, S. J. 1977. Inactivation of viruses during anaerobic sludge digestion. *Journal of the Water Pollution Control Federation* 49(7):1642-1651.
6. Foster, D. H., and Engelbrecht, R. S. 1973. Microbial hazards in dispersing of wastewater on soil, pp. 247-270. In *Recycling Treated Municipal Wastewater and Sludge through Forest and Cropland*, W. E. Sopper and L. T. Kardos, eds. University Park, Pennsylvania: Pennsylvania State University Press.
7. Briscoe, John. 1976. Public Health in Rural India. The Case of Excreta Disposal. Research Paper No. 12. Doctoral dissertation, Harvard University, Center for Population Studies, Cambridge, Massachusetts.
8. Müller, G. 1960. Tubercle bacteria present in sludge produced in mechanical and biological sewage treatment plants. Abstracted by W. J. Muller, *Sewage and Industrial Wastes* 32(9):1030.
9. Fitzgerald, P., and Ashley, R. F. 1973. Experimental studies on parasite survival in digested sludge. In *Parasitological Study of Sludge and the Effect of Some Human, Animal, and Plant Parameters. Final Report, 1971-1973*. Urbana, Illinois: University of Illinois (submitted to the Metropolitan Sanitary District of Greater Chicago).
10. Wilson, H. 1945. Some risks of transmission of disease during the treatment, disposal and utilization of sewage effluent and sewage sludge. *Sewage Works Journal* 17:1297-1298.
11. Wright, W. H.; Cram, E. B.; and Noland, W. D. 1942. Preliminary observations on the effect of sewage treatment processes on the ova and cysts of intestinal parasites. *Sewage Works Journal* 14:1274-1280.
12. Muller, W. 1954. Sludge treatment and pathogenic organisms. *Gesundheits-Ingenieur* 75:187.
13. Majumder, N.; Prakasam, T. B. S.; and Suryaprakasam, M. V. 1960. Critical study of septic tank performance in rural areas. *Journal of the Institute of Engineers* (India) 40:742-761.
14. Rudolfs, W.; Falk, L. L.; and Ragotzkie, R. A. 1950. Literature review on the occurrence and survival of enteric, pathogenic, and relative organisms in soil, water, sewage, and sludges, and on vegetation. I. Bacterial and viral diseases. *Sewage and Industrial Wastes* 22(10):1261-1281.
15. Dotson, G. K. 1973. Some constraints of spreading sewage sludge on croplands, pp. 67-80. In *Land Disposal of Municipal Effluents and Sludges: Proceedings of Conference on the Land Disposal of Municipal Effluents and Sludges, March 12-13, 1973, Rutgers University, New Brunswick, New Jersey*. EPA Publication EPA-902/9-73-001. Washington, D.C.: U.S. Environmental Protection Agency.

PART III

TECHNOLOGY OF ANAEROBIC FERMENTATION: ENGINEERING PROCESS DESIGN

Chapter 1

General Design
Considerations

Extensive work on biogas plant* design has taken place in several countries, particularly in India, but also in the United States, the People's Republic of China, Germany, and Taiwan, and several monographs have been published describing their design and construction.[1,2,3,4] These plants can be designed either to process a given amount of waste material, or to produce a given quantity of gas for a specific use or uses. If the objective is to process a certain amount of waste material generated from the wastes of a family and the cattle they own, then the digester size, and the gas produced, will be directly proportional to the waste material available. If a digester is to be designed to produce a given quantity of gas for specific uses—such as cooking, lighting, and running a small engine—the approach will be different. Based on the quantity of gas required, an estimate can be made of the amount of waste material needed, and thus the size of the gas plant can be determined. Although the approach is different, the limiting constraint in both cases is the quantity of waste material needed to produce the desired amount of gas.

Irrespective of the approach, an understanding of the parameters that govern the process of anaerobic digestion is essential in developing criteria for biogas plant designs.

In this chapter, and elsewhere in this report, the yield of methane from the anaerobic digestion of various raw materials is stated in a variety of ways— based on the weight of volatile solids added, volatile solids destroyed, dry weight of substrate, wet weight of substrate, etc. Unfortunately, the various sources quoted have not included enough information to permit reduction of all the yield data to a common basis. The reader is cautioned, therefore, to examine carefully the yield basis for any particular process or design before making yield comparisons and, above all, to be clear what input (loading) basis is to be used for any given design.

---

*Although technically anaerobic digesters, these installations are intended to produce methane and have become known as biogas plants—particularly when built on a village or single-farm scale. For this reason, they are referred to as biogas plants in this chapter.

Raw Materials

The collection and processing of the raw materials, and their storage and transportation to the digester, are obviously important considerations in the generation of methane from human, animal, and agricultural wastes. These problems have been discussed in general terms in Chapter 2 of the previous Part. However, since they are critical to the design of a biogas system, additional emphasis is warranted. The major options affecting the process design of a biogas installation are illustrated in Figure III-1.

Digesters can be constructed to process the wastes either on the site of their origin or on a site away from it. Economic justification for transporting wastes depends on the quantities of waste materials available, the distances involved, and the costs of transportation. For example, when designing a biogas plant for the production of methane from wastes generated by a family and its cattle, where a household latrine is used and the cattle are confined to a shed or an area adjacent to the house, it is logical to process them on-site. If, on the other hand, the plant is to be designed to process wastes from a block of houses or a village, it is worthwhile to consider processing the wastes off-site, that is, away from the residential area.

In the case of individual household biogas plants, the handling and processing of the wastes can be accomplished in two ways: 1) wastes such as cattle-shed wastes (dung, urine, straw), vegetable matter (crop residues, garbage), and night soil may be collected by the villager and transported manually to the plant, where they will be digested; and 2) if it is socially acceptable—and physically practical—the water closet or privy may be connected directly to the digester. (See discussion of use of latrines in Part II, Chapter 5.) As noted earlier, there are circumstances, as in the case of small farms and village dairy cooperatives, where

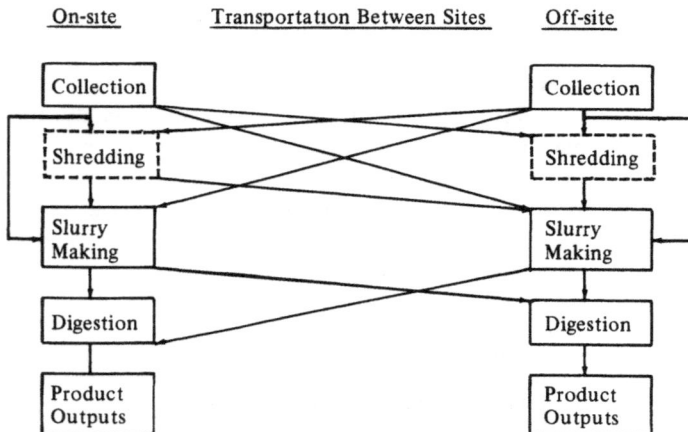

FIGURE III-1 Options in collection and processing of raw materials.

it may also be advisable and practical to design the biogas plant to receive animal wastes directly, as well. In this case, slatted floors would be installed in the cow shed (pig sty, poultry house, etc.) to permit the wastes to drop directly into an equilibration or mixing chamber (Figure III-2).

Where manure and agricultural wastes must be stored briefly before they are added to the digester, a storage shed should be provided. These sheds may be constructed of local materials such as bamboo, tree limbs, and palm thatch, or bricks, tiles, and mortar, depending on available material and financial resources. When the materials are fed to the digester, they can be transported by ox cart or other animal-drawn vehicle, or by shoulder bucket, for preparing the slurry. It is best to provide containers designed to safeguard the health of the workers—and any animals— transporting manure and, particularly, night soil. In many parts of Asia, and in developing countries elsewhere, large quantities of manure and night soil are handled manually for use as fertilizer, and it is not anticipated that their use for methane production would pose additional problems.

If aquatic vegetation is to be part of the raw material fed to the biogas plant, it can be collected, for example, by two or more persons pulling a net stretched between two bamboo poles through the water. Water hyacinth, algae, and other weeds may be dried in open sunshine for easier handling. The dried material may also be used as bedding in cow sheds to soak up the urine and then subsequently fed to the digester.

Whatever the mode of handling wastes, it should be geared to the human and animal labor available in the area.

Quantities of Gas Produced

The amount of methane produced by the anaerobic process depends primarily on the nature of the raw material.* Gas production also depends on

*The literature on methane generation by anaerobic digestion is far from consistent in assessing the quantity of material required to produce a given quantity. The most common practice among sanitary engineers is to base methane production on the quantity of volatile solids destroyed (see Glossary). However, this measurement requires facilities that are not always available in situations where biogas-generating experiments are carried out. Consequently, authors have reported methane production based variously on total (that is, dry) solids added, volatile solids added, and volatile solids destroyed. Some attempts have been made to provide the information needed to convert from one basis to another (see Ref. 14, this chapter, for example). However, the relationships vary with the specific materials used because the composition of waste materials is dependent on source—in the case of vegetable wastes, on the type of plant, soil conditions, fertilizers used, in the case of manures, the type of animal, the particular breed, grazing conditions, feed consumed, and state of health. (See Part II, Chapter 2, Raw Materials and Their Preparation.) A consistent basis that would permit intercomparisons of various materials and various advocates' claims of efficiency would indeed be useful; however, in many circumstances, such a consistency might well be misleading. Therefore, throughout this report, the panel has felt it best to report methane production just as it appears in the literature.

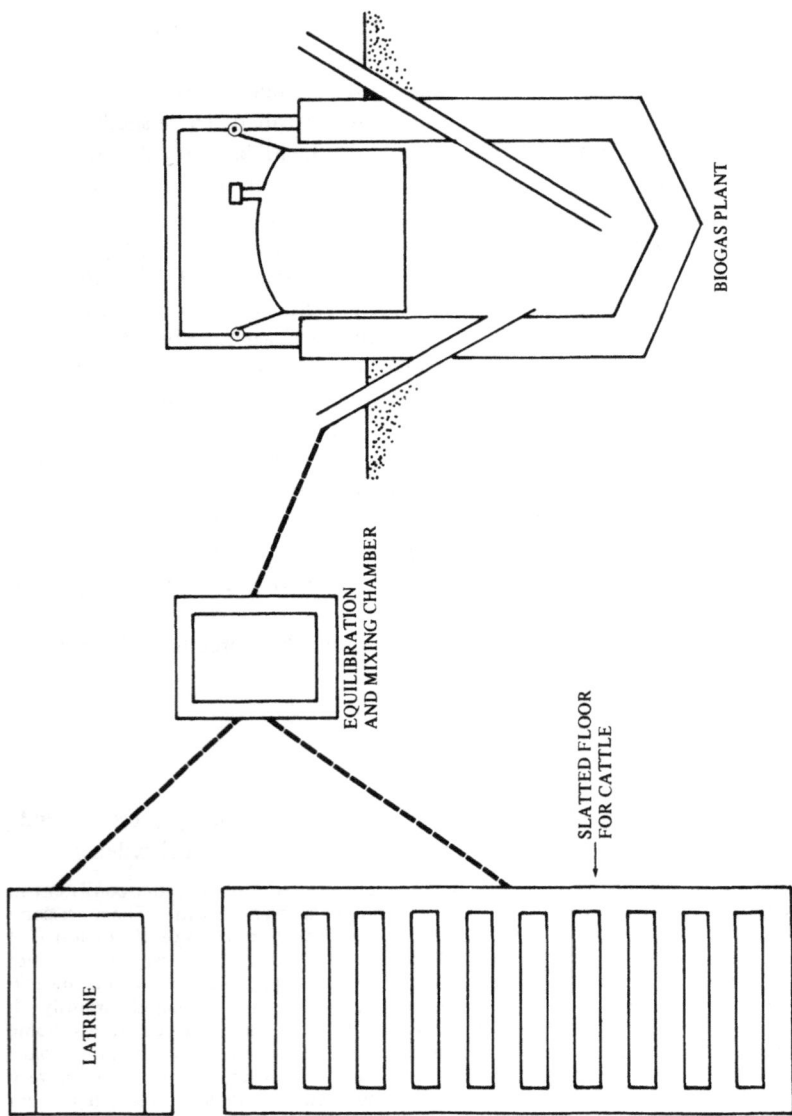

FIGURE III-2   Biogas plant for the generation of methane from night-soil and cattle-shed wastes.

the temperature and period of digestion; investigators have consistently observed that gas production is greater at higher temperatures, oth r factors being equal and no toxic substances being present. Chapter 3 of this Part and Chapter 1 of Part II contain a more detailed discussion of the role of temperature.

Influence of Quality of Raw Material on Gas Production

Various authors have reported on the quantity and composition of the biogas produced from different waste materials, and typical results are summarized in Table III-1.

When gas production data are evaluated, it should also be noted that the wastes from similar sources may yield entirely different quantities of gas because of differences in either the characteristics of manurial solids or operating conditions. Furthermore, the quantities of manure produced also differ because of the difference in feed ration; the cows in the Indian subcontinent, for instance, produce about one-third the quantity of manure produced by cows in the United States.[5] (See Part II, Chapter 2.)

It has also been observed that vegetable matter from young plants produces more gas than that from older plants and that dry vegetable matter produces more gas than green vegetable matter.[5] Brush and weeds that flourish in many developing countries during the monsoons may be harvested and permitted to dry, although not completely, during the intermittent periods of sunshine that occur throughout this season. The semidry material may then be chopped and stored in a storage shed, as mentioned earlier. Harvesting of young and succulent brush and weeds is relatively easy because of the looseness of the soil during wet weather. It also helps in the conservation of other vegetation and, to some extent, may reduce the incidence of harmful snakes and other pests.

Influence of Carbon-to-Nitrogen Ratio on Gas Production

Gas production has been found to vary significantly with the mixture of materials used. Experience has shown that gas production can be increased by supplementing substrates that have a high carbon content with substrates containing nitrogen, and vice versa. In the chapter on Raw Materials (Part I, Chapter 2), it was pointed out that although the concentration of nitrogen needed for anaerobic digestion is not very large, nevertheless, the ratio of carbon to nitrogen (present and available) in the raw materials is important for efficient production of methane. Under normal conditions, equal amounts of methane and carbon dioxide are produced in the digester. (See Part II, Chapter 1, Biological Mechanisms.) However, nitrogen is needed for incorporation into the cell structure; thus, if there is insufficient nitrogen

TABLE III-1 Yield of Biogas from Various Waste Materials[a]

Raw Materials	Biogas Production per Unit Weight of Dry Solids		Temperature		CH$_4$ Content in Gas (%)	Fermentation Time (Days)	Reference
	ft³/lb	m³/kg	°F	°C			
Cow dung	5 3	0.33	–	–	–	–	10
Cattle manure	5	0.31	–	–	–	–	3
Cattle manure (India)	3.6-8.0	0.23-0.50	52-88	11.1-31.1	–	–	11
Cattle manure (Germany)	3.1-4.7	0.20-0.29	60-63	15.5-17.3	–	–	11
Beef manure	13.7[b]	0.86[b]	95	34.6	58	10	12
Beef manure	17.7	1.11	95	34.6	57	10	12
Chicken manure	5.0[c]	0.31[c]	99	37.3	60	30	13
Poultry manure	7.3-8.6[f]	0.46-0.54[f]	90.5	32.6	58	10-15	14
Poultry manure	8.9[c]	0.56[c]	123	50.6	69	9	20
Swine manure[d] [e]	11.1-12.2	0.69-0.76	90.5	32.6	58-60	10-15	14
Swine manure[d] [e]	7.9	0.49	91	32.9	61	10	15
Swine manure	16.3	1.02	95	34.6	68	20	15
Sheep manure[d]	5.9-9.7	0.37-0.61	–	–	64	20	15
Forage leaves	8	0.5	–	–	–	29	16
Sugar beet leaves	8	0.5	–	–	–	14	16
Algae	5.1	0.32	113-122	45-50	–	11-20	21
Night soil	6	0.38	68-79	20-26.2	–	21	17

(a)Some figures have been rounded to the nearest tenth.
(b)Based on total solids.
(c)Based on volatile solids fed.
(d)Includes both feces and urine.
(e)Animals on growing and finishing rations.
(f)Based on volatile solids destroyed. On the basis of conversion efficiencies given in Ref. 13, these results may be expressed as 4.0-4.7 ft³/lb dry solids added, or 0.26-0.30 m³/kg.

present to permit the bacteria to reproduce themselves, the rate of gas production will be limited by the nitrogen availability. If, on the other hand, there is more nitrogen available than is needed to enable the cells to reproduce normally, ammonia will be formed. Its concentration may rise to the point where it inhibits further growth, and the production of methane will slow down or even cease.

In summary, if the C/N ratio is too high, the process is limited by the nitrogen availability, and if it is too low, ammonia may be found in quantities large enough to be toxic to the bacterial population.[7] It has been found, as a matter of practice, that it is important to maintain a C/N ratio (by weight) close to 30 to achieve an optimum rate of digestion.[2,8] It should be noted,

TABLE III-2 Approximate Nitrogen Content and C/N Ratios by Weight of Various Waste Materials[a] (Dry-Weight Basis)

Material	N (%)[b]	C/N
Animal Wastes		
Urine	15-18	0.8
Blood	10-14	3
Fish scraps	6.5-10	5.1[c]
Mixed slaughterhouse wastes	7-10	2
Poultry manure	6.3	—
Sheep manure	3.8	—
Pig manure	3.8	—
Horse manure	2.3	25[c]
Cow manure	1.7	18[c]
Farmyard manure (average)	2.15	14
Night Soil	5.5-6.5	6-10
Plant Wastes		
Young grass clippings (hay)	4.0	12
Grass clippings (average mixed)	2.4	19
Purslane	4.5	8
Amaranthus	3.6	11
Cocksfoot	2.6	19
Lucerne	2.4-3.0	16-20
Seaweed	1.9	19
Cut straw	1.1	48
Flax waste (phormium)	1.0	58
Wheat straw	0.3	128
Rotted sawdust	0.25	208
Raw sawdust	0.1	511
Household Wastes		
Raw garbage	2.2	25
Bread	2.1	—
Potato tops	1.5	25
Paper	nil	—

[a]Source: Gotaas (1956)[18] and Fry and Merrill (1973).[4]

[b]Total nitrogen.

[c]Nonlignin carbon.

however, that a substance resistant to microbial attack will not be digested, regardless of its C/N ratio. Peat, for example, cannot be mineralized* easily because it responds slowly to microbial attack, and although it has a low C/N ratio, ammonia will not be formed in significant quantities. In assessing the C/N ratios for optimizing methane production, it is important to consider not only the waste materials but also their relative ease of degradation.

Table III-2 shows the C/N ratios of a variety of raw waste materials. With the aid of the values presented in this table, waste materials that are low in carbon can be combined with materials high in nitrogen, and vice versa, to achieve the optimal C/N ratio of 30:1.[2,3,4]

Stimulating Effects of Various Materials

The effects on biogas production of adding small quantities of plant materials to cow dung are illustrated in Table III-3.

Construction Materials

Practical information on the construction of biogas plants in developing countries is available from reports of the long-term projects in India,[2,21,22,23] and in reports of experience in Taiwan[24] and Korea.[25]

The materials required for the construction of two types of biogas plants that produce about 100 ft^3 (3 m^3) of gas per day in temperate regions are listed in detail in Appendix 1. The designs are illustrated in Figures III-3 and III-4. The gas produced from these plants is reported to be adequate for the cooking needs of a family of five. The materials mentioned by the authors (see Appendix 1) should be available without difficulty in most developing countries. There is a difference in the quantity of materials required for the two approaches. With the increased demand and cost of construction materials, it is unlikely that a villager can afford to buy the 40 bags of cement suggested by Singh (Figure III-4). In rural India, and perhaps in other developing countries, cement is in short supply and substitutes such as lime mortar with waterproofing should be considered for the construction of the digester. Where feasible, glazed pottery rings may be placed over one another and used as substitutes for brick and mortar, or concrete. Where these rings are used they must be cemented together and the joints checked for leaks. In villages in developing countries, the manufacture of these rings should not pose a serious problem since the raw materials and labor required for fabrication are generally available.

Construction materials such as slag, bricks, crushed stone, lime, mortar, asbestos, galvanized pipes needed for inlets and outlets, manhole covers, and

*i.e., have its nutrient elements converted to soluble or semisoluble mineral salts that can be transported through the soil by water and can be absorbed and utilized by plants.

TABLE III-3. Biogas Production from Anaerobic Fermentation, at Room Temperature, of Mixtures of Cow Dung and Agricultural Wastes[a]

| | Biogas Production per Unit Weight of Dry Solids | | | | Gas Composition after 21 days[b] (%) | | |
| | At the end of 24 days | | At the end of 80 days | | | | |
Raw Material	ft^3/lb	m^3/kg	ft^3/lb	m^3/kg	CH_4	H_2	CO_2
1. Cow dung	1.0	.063	3.3	.21	60.0	1.1	34.4
2. Cow dung + 0.4% cane sugar	1.1	.070	3.3	.21	57.6	2.1	38.4
3. Cow dung + 1% ashes	.98	.061	3.0	.19	60.4	2.9	34.4
4. Cow dung + 2.4% fresh leguminous leaves (25% dry matter, 2.31% N)	1.0	.063	3.2	.20	61.6	4.0	32.0
5. Cow dung + 1.2% sarson oil cake (94% dry matter, 4.74% N)	1.0	.063	3.3	.20	67.7	–	30.4
6. Cow dung + 1% cellulose	1.3	.084	3.3	.21	52.8	–	44.0
7. Cow dung + 0.4% casein (12.6% N)	1.4	.087	3.5	.22	64.0	2.4	32.0
8. Cow dung + 1% cane sugar + 1% urea (44.5% N)	1.4	.087	4.2	.26	68.0	–	30.6
9. Cow dung + 1% cane sugar + 1% $CaCO_3$	1.5	.091	3.9	.24	70.0	–	28.0
10. Cow dung + urine (4% solids) at 20 ml/100 g (15 fl. oz/lb)	1.4	.087	3.9	.24	67.0	–	32.0
11. Cow dung + 0.4% charcoal	1.0	.065	2.6	.16	65.6	–	32.0
12. Cow dung + 20% dry, non-leguminous leaves (1.71% N)	1.3	.081	3.5	.22	68.0	0.6	28.0

[a] Adapted from Laura and Idnani (1971).[17] Because the fermentation was carried out at room temperature (approximately 70°F or 21°C), data are included for a period of 80 days, far beyond the normal 2- to 4-week detention time characteristic of the data in Table III-1.

[b] The composition of the gas at the end of the 80-day period was not reported. Presumably, the CH_4/CO_2 ratio would not have changed significantly in that interval.

elbows and bends are generally available. Therefore, their availability should not be a limiting factor for the construction of biogas plants. The single most costly item is the gas holder, which may be a drum constructed of mild steel sheets.

In addition to digesters constructed with brick, cement, concrete, and steel, other types can be built. For example, a small digester made from an oil drum, using an innertube as the gas holder, has been built in prototype form and tested successfully in Mexico by Jedlicka.[26] (However, one digester of this type will not produce enough biogas for a family.) In countries where "traditional" biogas plants exist (i.e., bricks, mortar, or steel), however, the reluctance to adopt new methods may interfere with initial acceptance of new designs.

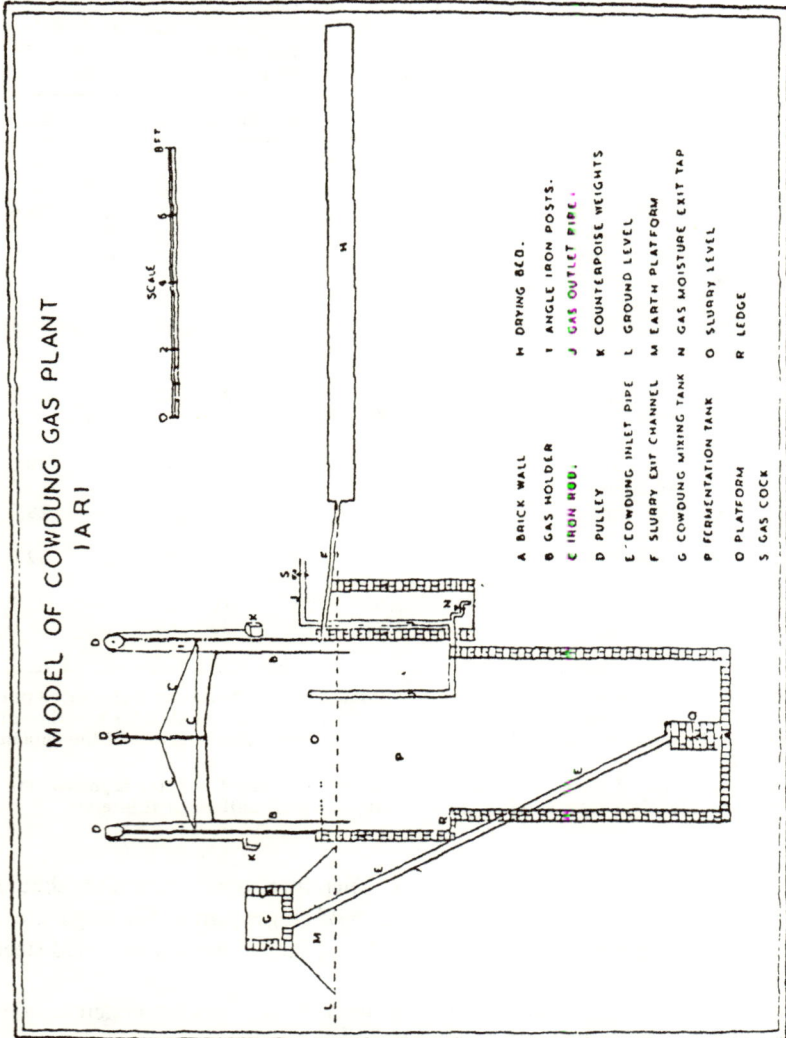

MODEL OF COWDUNG GAS PLANT
I A R I

SCALE

A BRICK WALL	H DRYING BED.
B GAS HOLDER	I ANGLE IRON POSTS.
C IRON ROD.	J GAS OUTLET PIPE.
D PULLEY	K COUNTERPOISE WEIGHTS
E COWDUNG INLET PIPE	L GROUND LEVEL
F SLURRY EXIT CHANNEL	M EARTH PLATFORM
G COWDUNG MIXING TANK	N GAS MOISTURE EXIT TAP
P FERMENTATION TANK	O SLURRY LEVEL
O PLATFORM	R LEDGE
S GAS COCK	

FIGURE III-3   Biogas plant designed by Acharya, developed at Indian Agricultural Research Institute.[21]

70

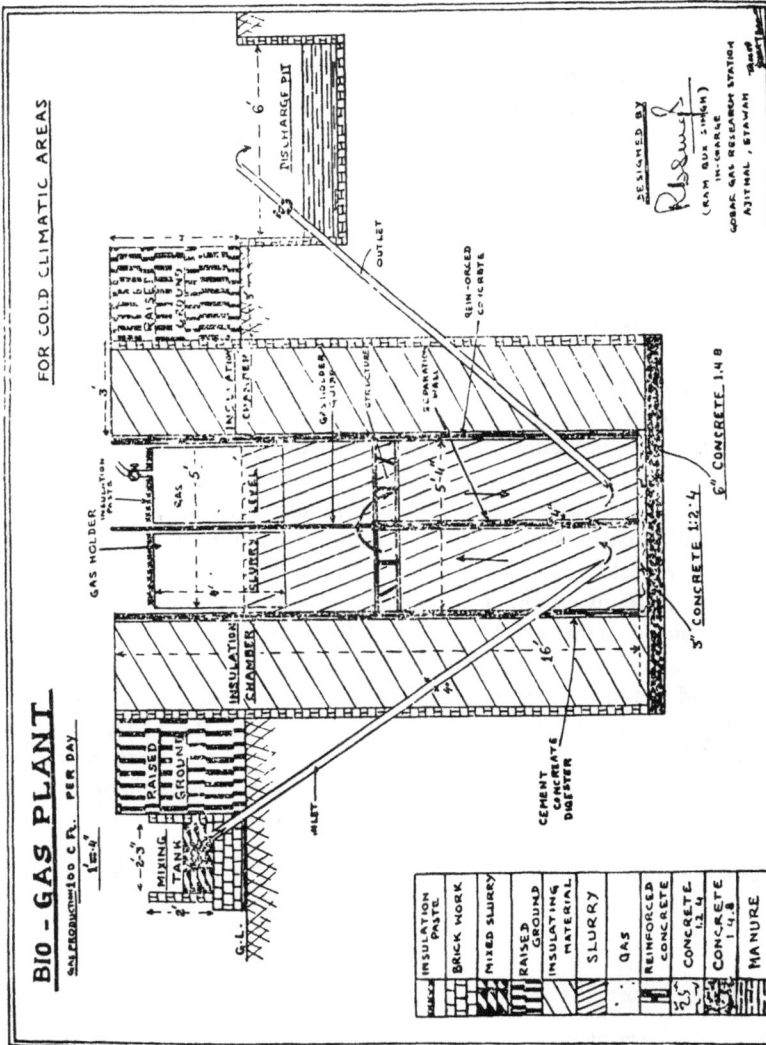

FIGURE III-4 Biogas plant designed by Singh, developed at Gobar Gas Research Station. [2]

71

The horizontal displacement type of digester pioneered by Fry is claimed to be economical and easy to install and operate.[3] Materials for construction of this type of digester may include: 1) a long cylindrical metal drum or concrete pipe lying horizontally on (or partially in) the ground; or 2) a long rectangular tank with a (horizontal) cylindrical roof. The tank and the roof may be constructed of concrete, cinder, or slag. However, prefabricated structures such as drums and cylindrical concrete pipes can be transported to the site where the gas plant will be constructed. Although this type of digester has some interesting features, its applicability and performance have yet to be tested in developing countries; vertical digesters are the rule in countries with biogas plants. The materials needed for construction of the horizontal type of digester are similar to those for a vertical digester and so are also available in most developing countries.

As the size of the digester increases, more materials will be needed. In larger digesters (500 ft^3 or 15 m^3 gas/day), where mixing and heating of the digester contents are required, additional equipment such as pumps, mixers, heating coils, and water boilers will be needed. In developing countries, particularly in rural areas, obtaining such equipment may be difficult.

Labor for the construction of biogas plants in developing countries is usually available. When large-size community digesters are to be built for villages, skilled personnel are needed to maintain these units. Without a government policy decision to provide the technical skill needed to operate and maintain the large digesters, it is unlikely that methane generation using community biogas plants will be accepted. In India, for example, the Khadi and Village Industries Commission provides incentives in the form of financial assistance and technical guidance for the construction and maintenance of "Gobar"* gas plants.[23] Furthermore, in some countries there may be many people available with sufficient education or technical aptitude to be trained to supervise the construction and maintenance of the biogas plants if a national program to construct such plants is implemented.[27]

*"Gobar" is the Hindi word for dung.

References

1. Unpublished collection of 66 titles of articles on-biogas published from 1932-1974. n.d. Darmstadt, Germany: Kuratorium für Technik und Bauwesen in der Landwirtschaft (KTBL) (mimeographed).
2. Singh, Ram Bux. 1971. *Bio-gas Plant: Generating Methane from Organic Wastes*. Ajitmal, Etawah (U.P.), India: Gobar Gas Research Station.
3. Fry, L. J. 1974. *Practical Building of Methane Power Plants for Rural Energy Independence*. Santa Barbara, California: Standard Printing.
4. Fry, L. J., and Merrill, R. 1973. *Methane Digesters for Fuel Gas and Fertilizer*. Newsletter No. 3. Santa Cruz, California: New Alchemy Institute.

5. Sathianathan, M. A. 1975. *Bio-gas Achievements and Challenges*. New Delhi: Association of Voluntary Agencies for Rural Development.
6. Boshoff, W. H. 1967. Reaction velocity constants for batch methane generation on farms. *Journal of Agricultural Science* 68:347.
7. Harmson, G. W., and Kolenbranden, G. J. 1965. Soil inorganic nitrogen, pp. 43-92. In *Soil Nitrogen*, W. V. Bartholomew and F. E. Clark, eds., No. 10. Madison, Wisconsin: American Society of Agronomy.
8. Acharya, C. N. 1958. *Preparation of Fuel Gas and Manure by Anaerobic Fermentation of Organic Materials*. Indian Council of Agricultural Research Series, Bulletin No. 15. New Delhi: Indian Council of Agricultural Research.
9. Desai, S. V., and Biswas, S. C. 1945. Manure and gas productions by anaerobic fermentation of organic wastes. *Indian Farming* 6(2):67-71.
10. Joppich, W. 1957. German farms too use fuel gas plants. *Indian Farming* 6(11):35-40.
11. Loehr, R. C., and Agnew, R. W. 1967. Cattle wastes—pollution and potential treatment. *American Society of Civil Engineers. Sanitary Engineering Division Journal* 93:55-72.
12. Klein, S. 1972. Anaerobic digestion of solid wastes. *Compost Science* 13(Jan.-Feb.):6-16.
13. Gramms, L. C.; Polkowski, L. B.; and Witzel, S. A. 1971. Anaerobic digestion of farm animal wastes (dairy bull, swine and poultry). *ASAE Transactions (American Society of Agricultural Engineers)* 14:7-11, 13.
14. Jeffrey, E. A.; Blackman, W. C.; and Ricketts, R. L. 1964. *Aerobic and Anaerobic Digestion Characteristics of Livestock Wastes*. University of Missouri Engineering Series Bulletin 57. Columbia. University of Missouri.
15. Reinhold, F. n.d. Gasleistung verschiedener Ausgangsstoffe (mimeographed).
16. Mudri, S. S. 1967. Some observations on the anaerobic digestion of night soil. *Environmental Health* (India) 9:133-136.
17. Laura, R. D., and Idnani, M. A. 1971. Increased production of biogas from cow dung by adding other agricultural waste materials. *Journal of the Science of Food and Agriculture* (India) 22:164-167.
18. Gotaas, Harold B. 1956. *Composting*. Geneva, Switzerland: World Health Organization.
19. Savery, C. William, and Cruzan, Daniel C. 1972. Methane recovery from chicken manure digestion. *Journal of the Water Pollution Control Federation* 44:2349-2354.
20. Golueke, C. B., and Oswald, W. 1964. Power from solar energy via algae-produced methane. *Solar Energy* 7:86-92.
21. Acharya, C. N. 1956. Your home needs a gas plant. *Indian Farming* 6(2):27-30.
22. Kulkarni, D. V. 1963. *Gobara Gyasplanta* [Cow-dung Gas Plant]. Faridabad, Haryana, India: Pradip Publications. (In the Kannada language.)
23. Khadi and Village Industries Commission. n.d. *Gobar Gas—Why and How*, p. 20. Bombay: Khadi and Village Industries Commission.
24. Yu, Ju-tung. 1965. *Notes on Raising Pigs to Gain Wealth*. Taipei: Feng Nien She.
25. Kim, S. G., and Libby, L. W. 1972. Rural infrastructure, pp. 69-75. In *Korean Agricultural Sector Study Special Report No. 2*. East Lansing: Michigan State University, Department of Agricultural Economics.
26. Jedlicka, Allen D. 1974. Comments on the Introduction of Methane (Bio-gas) Generation in Mexico with an Emphasis on the Diffusion of Back-yard Generators for Use by Peasant Farmers. Cedar Falls, Iowa: University of Northern Iowa, College of Business and Behavioral Sciences. (Unpublished paper.)
27. Patankar, G. L. 1974. Role of gobar gas plants in agro-industries. *Khadi Gramodyog: Journal of Rural Economy* (India) 20(April):351-357.

Chapter 2

Design Criteria

The design of biogas plants for rural areas of developing countries can be based on either of two objectives: 1) the production of biogas from digestible materials available at the site, with use of the gas determined by the amount produced; and 2) the production of a required quantity of biogas for a specific purpose such as cooking, lighting, or operating an internal-combustion engine. In general, it is preferable to design a plant with the second objective in mind, for it is easier to vary the fraction of available waste materials fed to a plant than to have insufficient gas to operate devices designed to use the product.

To justify the construction of a gas plant, three prerequisites must be met: 1) the smallest gas plant that can be justified economically should be able to produce enough cooking gas for a family of four when efficient burners are used, that is, $12-15\,\text{ft}^3$ ($0.34-0.42\,\text{m}^3$) per person per day of gas with methane content of 60 percent or more; 2) the quantity of available waste materials must be at least enough to meet this minimum production goal, which will require, for a household of five persons, the dung from at least six cattle (each animal producing approximately 20 lb, or 10 kg, of wet dung daily) or the night soil from 60 adults;* and 3) there should be enough water available to make a slurry of the raw materials fed to the digester. In the case of cow dung, one part of water is required for every part of dung (by volume). In the case of night soil, flush-water per person per day should be about a liter, since excessive dilution is not conducive to methane fermentation.

Subsidiary considerations involve the on-site/off-site options outlined in Figure III-1. That is: 1) adequate space for the biogas plant must be available close to the stable where the animals are housed, or to the latrines and the place where the gas will be used; 2) space must be provided adjacent to the gas plant to dry, store, or compost the digested sludge from the biogas plant;**

---

*It should be stressed that it is the manure-loading rate, not the number of animals, that dictates the design of the biogas plant. The statement in the text is based on an "average" production rate. (See Part II, Chapter 2, Materials of Animal Origin.)

**Although there is a risk of some loss of nitrogen during composting of the sludge, in some situations composting may be needed to kill off surviving pathogens or parasites. (See Part II, Chapter 4, for discussion of nitrogen balance.)

and 3) the gas plant must not be constructed within 50 ft (15 m) of a drinking-water well.[1]

The availability of raw materials and labor and the assurance of technical supervision, when needed, for operation of the plant must also be taken into account in the design, as well as the various parameters that influence the rate of gas production, mentioned in previous chapters.

Design Considerations Dependent on Size

Size-dependent considerations for the design of a biogas plant include:

- The amount of raw material available;
- The type of raw material available;
- The average particle size of the raw material;
- Heating requirements;
- Mixing requirements; and
- Construction materials available.

All factors except the last are important when determining the amount of gas produced and its methane content. Under optimum environmental conditions for digestion, the amount of gas produced is proportional to the amount of waste materials used. The effect of the qualitative nature and C/N ratio of the waste materials on the rate of gas production has already been discussed. Materials that can be degraded easily will be stabilized sooner than more resistant materials. Hence, for production of the same amount of gas, detention times will be shorter and digester size smaller when easily degraded materials are used as feed. The particle size of the raw materials also affects the amount of gas produced; materials shredded into small pieces will ferment better and give fewer problems than unchopped and bulky materials. It is advisable to shred into small pieces materials such as straw, wood chips, leaves, and bagasse before feeding them to the digester. This increases gas production by increasing the surface area exposed to bacterial attack, thereby reducing the detention time in the digester. Moreover, it enables the slurry to flow smoothly to the digester, with less scum production. Problems due to clogging of inlet and outlet pipes may also be minimized by shredding the digester feed materials.

If it is possible to heat the digester, thus accelerating the digestion process, the detention time will be reduced and the digester size can be smaller than for an unheated unit. Since the size of the digester is determined by the volume of the slurry formed (from the amount of raw materials needed daily for a given gas production) multiplied by the detention time, reduction of the time directly reduces the minimum size required. One should keep in mind, however, that energy is required to heat a digester. Unless this energy has a trade-off in terms of net gas production, reduction in digester size, and process stability, heating of a digester cannot be justified. In many instances, however, the heating may well be economical.

The amount of mixing influences the size of the digester in somewhat the same manner as particle size, in that it exposes new surfaces to bacterial action, and prevents slowdown of bacterial activity caused by local depletion of nutrients or concentrations of metabolic products. Generally, it is not necessary to agitate the contents of small digesters when low gas production rates are satisfactory because of the long detention times (50–60 days). However, in larger gas plants the size of the digester can be reduced considerably by incorporating an agitator in the design.

Finally, a lack of construction materials may limit the size of the biogas plant, even though there may be an adequate supply of agricultural, animal, and human wastes to generate the required gas.

Design Considerations Independent of Size

Irrespective of the size of the digester, certain choices must be made to ensure its optimal performance. These involve:

- Minimizing corrosion problems;
- Preventing contamination of drinking-water sources;
- Determining the best flow of materials (continuous, semicontinuous or batch); and
- Selecting construction materials.

Corrosion is a serious problem because of the constant exposure of the metal parts of the system to the hydrogen sulfide and the organic acids that may either be present or formed in the digestion process. Thus, it is necessary to paint these parts with corrosion-resistant paint.

It was pointed out in Part II, Chapter 5, that groundwater drinking supplies that are protected from surface runoff are normally not endangered by the use of digested sludge or the storage of manure, because of the filtration capability of 3 feet or more of soil. It should be emphasized here, therefore, that the engineering process design for the method of handling the raw material and the digested slurry must make provisions for adequate separation of these materials—and possible runoff—from drinking-water supplies.

The flow of materials must be considered before the system can be designed. In this respect, biogas plants for rural areas in developing countries may be broadly characterized as follows:

```
                              Biogas Plants
                                   |
        ┌──────────────────────────┴──────────────────────────┐
   Batch-fed (periodic)                              Daily-fed (semicontinuous)
        |                                                       |
 ┌──────┴──────────┐                          ┌────────────────┴────────────────┐
vertical          horizontal              vertical                    horizontal
                  displacement            (completely                 displacement
                  (plug flow)             mixed)                      (plug flow)
    |
 ┌──┴──────────┐
completely    not mixed
mixed         (if too small)
```

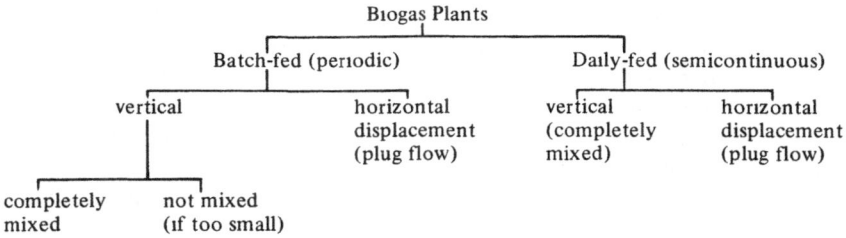

Batch-fed plants can be constructed where daily supplies of raw waste materials are difficult to obtain. Singh[2] suggests that batch-fed digesters be used to ferment coarse vegetable wastes, such as corn stalks or cobs and sugarcane husk, which cannot flow smoothly through a semicontinuously fed digester. Batch-fed digesters are charged with the waste, which is covered and permitted to ferment. After a start-up interval of about 2 weeks, gas production begins and continues for about 3 months. When gas production ceases, the digesters are opened, cleaned, and the slurry is disposed of on cropland as fertilizer. With a batch system, it is desirable to have at least two digesters so that one (or more) can always be in operation. However, periodic emptying of the digesters involves much labor and can be unpleasant.

Competely mixed and semicontinuously fed vertical digesters, with or without partitioning, have been popular in India, France, and Germany because of their reliability. Although vertical digesters that are completely mixed and fed daily on a semicontinuous basis with sewage sludge are known to perform efficiently, such efficiencies may not be achievable in digesters designed for biogas production in rural areas using agricultural-type materials. This is mainly because of the difficulty in maintaining the careful control needed and the vagaries of composition of the daily feed material.

Fry[3] reported that the horizontal displacement (plug-flow) type of digester, which he pioneered, is less prone than the vertical digester to problems such as scum formation.

Generally, in the design of biogas plants consideration is also given to the use of one or more separate compartments in a digester or to two digesters operated in a series. This makes small plants more expensive and may be unnecessary, but the advantages of having compartments in the digestion tank, or of having a series type of operation, are a higher degree of waste stabilization and increased gas production. Detailed discussion of compartmentation and series operation is presented in the monographs written by Singh[2] and Fry.[3]

Digesters of any type can be constructed either above or below ground. In temperate and tropical countries, when digesters are constructed above ground, painting them black will increase the absorption of solar energy to heat the digester.

The construction of various types of digesters using different construction materials has been described elsewhere.[4] The important thing is that durable building materials that require minimum maintenance should be specified. The use of nonconventional materials—such as tires, Plexiglas, and butyl rubber—for the construction of gas plants is risky because information is scarce on their suitability, performance, maintenance, and durability in rural areas. Furthermore, many of these materials are not available or are too costly in developing countries. Digesters made of concrete, stone, brick, and steel are durable; however, the high cost of steel and the difficulty of on-site fabrication may limit their construction in rural areas.

Comparison of Biogas Plants in a Developing and an Industrialized Country

Joppich[5] has compared the performance of three types of plants designed in Germany with prototype designs developed by the Indian Agricultural Research Institute (IARI) in New Delhi, India. The German plants were the Schmidt-Eggersgluss, Weber, and Kronseder plants. The first two are heated two-stage digesters equipped with separate gas holders, having mechanical mixing, and designed for operation on large farms. The third, or Kronseder plant, has a simpler design meant for the production of smaller quantities of gas such as may be used for cooking on small farms. In these plants, a semicircular tub is inverted and floated in a concrete cesspool, which receives manure, excreta, and other wastes. The gas that is evolved is collected in the tub. The IARI plants used in this study were unheated vertical digesters treating cow dung and equipped with gas holders. The results of this comparative study are given in Appendix 3.

The digester volume required for an animal unit in the German designs was about twice that of the Indian design, presumably due to a high content of straw in the German agricultural waste. The quality of gas produced per animal unit in the German digesters was better, since considerable amounts of fermentable materials other than cow dung were present in the digester feed. However, the quantity of gas produced in the Indian digester per pound of material fermented was higher than for the German unit. This was presumably due to the presence of more readily fermentable materials in the feed to the Indian digesters; the feed to the German digesters contained a higher percentage of straw, which is more difficult to digest. It is interesting that the gas produced in the German digester had a higher content of methane, indicating that a better C/N ratio may have existed in the agricultural wastes of Germany.

The results of this comparison point up the necessity for evaluating the characteristics of the indigenous raw wastes in designing biogas plants. In countries where data on the characteristics of indigenous waste materials are not available, it is worthwhile to perform pilot-plant studies before the design of projected biogas plants is undertaken.

Design Example

The design of a typical biogas plant will be given here, step by step, and in a list of materials given for the 100-ft^3/day plant common in India * on page 111.

We will begin with an assumed requirement of 100 kWh of electricity daily for a rural area of India, to be generated by using biogas from a cow-dung digester. The maintenance and operation of the proposed system will be handled by personnel specially trained for the purpose.

The assumptions on which the design is to be based are:

- The biogas will be used to fuel an engine-generator to produce the needed electrical power;
- The thermal efficiency of the engine is 25 percent;
- The electrical-mechanical efficiency of the generator is 80 percent;
- 1 ft^3 of biogas with methane content of 55 percent will provide 550 Btu (higher heat value if the methane content is greater);
- 50 percent of the fuel energy is transferable to cooling water;
- 1 kWh = 3,415 Btu;
- 10 lb of dung, plus the water needed to make up the slurry, occupies a volume of 2/7 ft^3 (1.8 l/kg of dung); and
- The digester will be heated to 85°F (29°C).

The quantities to be determined are:

- The amount of biogas needed to produce the required electrical output;
- The amount of cow dung required per day to generate the biogas;
- The dimensions of the digester and gas holder needed;
- The heating requirements of the digester [it will also be necessary to show whether the waste heat from the engine will be sufficient to operate the digester at 85°F (29°C)] ; and
- The weight of the output slurry removed each day.

The calculations will be in English units; results will be given in both English and metric units.

*It would be helpful for the reader to review Figure I-2 and the discussion in Part I of the criteria for determining design parameters, before beginning this section.

1. Biogas Needed

$$\text{Btu needed} = 3{,}415 \frac{\text{Btu}}{\text{kWh}} \times 100 \frac{\text{kWh}}{\text{day}} = 341{,}500 \text{ Btu/day.}$$

Since the thermal efficiency of the engine is only 25 percent, the actual heat required daily will be $341{,}500 \div 0.25 = 1{,}366{,}000$ Btu. With a generator efficiency of 80 percent, this means an overall daily heat requirement of $1{,}366{,}000 \div 0.8 = 1{,}707{,}500$ Btu. At 550 Btu/ft^3 of biogas, the daily biogas requirement is:

$$1{,}707{,}500 \frac{\text{Btu}}{\text{day}} \times \frac{1}{550 \frac{\text{Btu}}{\text{ft}^3}} = 3{,}105 \text{ ft}^3/\text{day, say}$$

$$3{,}100 \text{ ft}^3/\text{day or } 90 \text{ m}^3/\text{day.}$$

2. Quantity of Cow Dung Needed

Assume 1 lb of cow dung (20–25 percent dry solids) yields 1 ft^3 of gas. Therefore, we need 3,100 lb (1,400 kg) of dung per day. Assume dung production @ 25 lb (11 kg) per cow. Therefore, the number of cows needed to produce this quantity of dung is $3{,}100 \div 25 \cong 125$ cows, or 1.25 cows are equivalent to 1 kWh of power in terms of methane generation from the dung. In rural India, on the average, there is one cow for every two people. Therefore, in a village of 250 people, the potential exists for generating the 100 kWh of electricity.

3. Digester Volume and Dimensions

The digester must receive 3,100 lb of dung daily and must retain its charge for 50 days in order to generate the required amount of gas.

Since 10 lb of dung, when slurried, occupies 2/7 ft^3, the 3,100 lb of dung fed to the digester daily will occupy

$$2/7 \text{ ft}^3 \times \frac{1}{10 \text{ lb}} \times 3{,}100 \text{ lb} = 90 \text{ ft}^3 \text{ (2.5 m}^3\text{)}.$$

For a 50-day detention time the volume required is 90 ft^3/day × 50 days, or 4,500 ft^3 (127.4 m^3).

If a cylindrical digester is chosen, with a ratio of 1:1 for diameter to depth, the approximate dimensions will be:

Depth = 18′ (5.5 m)
Diameter = 18′ (5.5 m).

However, if a truncated conical (hopper) bottom is desired in the digester, the dimensions will be approximately as follows:

Diameter at surface	$\cong$ 19' (5.8 m)
Depth of the cylindrical portion	$\cong$ 13' (4.0 m)
Depth of the truncated hopper bottom	$\cong$ 6.5' (1.8 m)
Bottom diameter of the hopper	$\cong$ 6.5' (1.8 m).

4. Gas Holder Volume and Dimensions

Since the gas will be used on a regular basis and will be withdrawn at a relatively constant rate, the gas holder need have only half the volume of the required daily production. Thus, for a daily production of 3,100 ft^3 (88 m^3), the gas holder need have a capacity of only 1,550 ft^3 (44 m^3). For a cylindrical gas holder to fit into the top of the digester whose dimensions were just determined, a suitable diameter would be 18.5 ft (5.6 m), or 6 in. (15 cm) less than the diameter of the digester compartment. The height of the gas holder will then be:

$$\frac{\text{Volume}}{\text{Area}} = \frac{1,550 \text{ ft}^3}{\pi \times (9.25 \text{ ft})^2} = 5.8 \text{ ft (say 6 ft or 1.8 m)}.$$

Note that the weight of the gas holder not compensated by counterweights will compress the gas somewhat so that the actual volume of gas above the surface of the slurry will be less than the design volume.

5. Heat Requirement

The maximum heat requirement is calculated for the coldest month of the year. Since the geographical regions involved are tropical and subtropical, the minimum mean slurry temperature, if no heat is added, will be assumed to be 55°F ($\cong$ 13°C), and the minimum mean air temperature will be assumed to be 35°F (1°C).[2]

Heat Needed to Reach Operating Temperature. The operating temperature for the digester will be 85°F (29-30°C). With 3,100 lb of dung plus an equal weight of water added to the digester daily, the total incremental load is 6,200 lb (2,818 kg). If water and dung are assumed to have the same specific heat, the heat required to raise the temperature of the input slurry from 55°F to 85°F is

$$6,200 \text{ lb} \times \frac{1 \text{ Btu}}{\text{lb} \times \text{°F}} \times (85 - 55)\text{°F} = 186,000 \text{ Btu (19.6 kJ) per day}.$$

Heat Losses. Heat losses from the digester depend on the temperature of the surroundings. For the coldest month of the year, the following assumptions will be made:[2]

Mean air temperature: 35°F (1°C)
Mean temperature of earth next to digester wall: 57.2°F (14°C).

The "q" factors (heat conductivities) are assumed to be:

For the walls—concrete surrounded by dry earth below grade (or embankments)

$$q_w = 0.12 \text{ Btu/hr/ft}^2/°F \ (2.45 \text{ kJ/hr/m}^2/°C).$$

For the floor—concrete in contact with moist earth

$$q_f = 0.15 \text{ Btu/hr/ft}^2/°F \ (3.07 \text{ kJ/hr/m}^2/°C).$$

For the roof (i.e., top of gas holder)—14–16 gauge mild steel or galvanized iron sheets, insulated with 1-in. insulating board

$$q_r = 0.16 \text{ Btu/hr/ft}^2/°F \ (3.27 \text{ kJ/hr/m}^2/°C).$$

The area of the surfaces in contact with the surroundings, for the digester with the hopper bottom, are:

Wall area $= \pi \times 19 \text{ ft} \times 13 \text{ ft} = 776 \text{ ft}^2 \ (72 \text{ m}^2)$
Floor area $= \pi \times (9.5 \text{ ft} + 3.25 \text{ ft}) \sqrt{(6.5 \text{ ft})^2 + (9.5 \text{ ft} - 3.25 \text{ ft})^2}$
            $+ \pi \times (3.25)^2 = 394 \text{ ft}^2 \ (37 \text{ m}^2)$
Roof area $= \pi \times (9.25 \text{ ft})^2$
             $= 269 \text{ ft}^2 \ (25 \text{ m}^2).$

The heat losses are then calculated as follows:

Heat loss from the walls:

$$0.12 \frac{\text{Btu}}{\text{hr} \cdot \text{ft}^2 \cdot °F} \times 776 \text{ ft}^2 \times (85°F - 57°F) = 2,607 \text{ Btu/hr} \ (2,750 \text{ kJ/hr})$$

Heat loss from the floor:

$$0.15 \frac{\text{Btu}}{\text{hr} \cdot \text{ft}^2 \cdot °F} \times 394 \text{ ft}^2 \times (85°F - 57°F) = 1,655 \text{ Btu/hr} \ (1,746 \text{ kJ/hr})$$

Heat loss from the roof:

$$0.16 \frac{\text{Btu}}{\text{hr} \cdot \text{ft}^2 \cdot °F} \times 269 \text{ ft}^2 \times (85°F - 35°F) = 2,152 \text{ Btu/hr} \ (2,270 \text{ kJ/hr}).$$

The total daily heat loss for this example is then

$$24(2,607 + 1,655 + 2,152) = 153,936 \text{ Btu } (162,402 \text{ kJ}).$$

Finally, the total heat requirement is the sum of the heat losses at the operating temperature and the heat required to raise the slurry to the operating temperature:

$$153,936 + 186,000 = 339,936 \text{ Btu/day } (358,632 \text{ kJ/day}).$$

Heat Available from the Engine-Generator. The actual amount of heat produced by the engine-generator (i.e., not converted to electricity), based on an overall efficiency of 20 percent (80 percent × 25 percent), is 1,707,500–341,500 = 1,366,000 Btu (1,441,130 kJ) daily. (This is the same as 80 percent of 1,707,500 Btu.) With a total heat requirement of 339,936 Btu/day (358,632 kJ/day), this means that the amount of heat available for transfer to cooling water and for dissipation will be 1,366,000–339,936 = 1,026,064 Btu/day (1,082,498 kJ/day).

Fifty percent of this quantity is assumed to be potentially available for transfer to the cooling water; this means that 513,032 Btu (541,249 kJ) will be available per day to heat the cooling water. Since this quantity of heat is more than the total amount of heat required to maintain the slurry at the operating temperature (339,936 Btu, or 358,632 kJ daily), it will be feasible to heat the digester with "waste" heat from the engine-generator.

6. Weight of Slurry Withdrawn from the Digester

The total weight of slurry added daily to the digester was calculated to be 6,200 lb (3,100 lb manure + 3,100 lb water). During digestion 10 percent of the total slurry weight is volatilized.[2] Therefore the net weight of the slurry withdrawn per day is 6,200 × 0.9 = 5,580 lb (2,536 kg).

References

1. Mohanrao, G. J. 1974. Scientific aspects of cow dung digestion. *Khadi Gramodyog: Journal of Rural Economy* (India) 20(April):340-347.
2. Singh, R. B. 1971. *Bio-gas Plant: Generating Methane from Organic Wastes.* Ajitmal, Etawah (U.P.), India: Gobar Gas Research Station.
3. Fry, L. J. 1974. *Practical Building of Methane Power Plants for Rural Energy Independence.* Santa Barbara, California: Standard Printing.
4. Acharya, C. N. 1956. Your home needs a gas plant. *Indian Farming* 6(2):27-30.
5. Joppich, W. 1957. German farms too use fuel gas plants. *Indian Farming* 6(11):35-40.

Chapter 3

Operation and Maintenance

For many years, anaerobic digestion of waste materials has been taking place, unattended, in a variety of natural and man-made situations. Nevertheless, when production of methane is the goal, a certain amount of care and attention is necessary to operate a biogas plant and maintain its production at a level that makes its construction and operation economically worthwhile.

A healthy anaerobic digestion process requires optimum conditions for the system environmental-control parameters and other characteristics. These are temperature, anaerobic conditions, nutrients in adequate concentrations, pH within tolerable limits, and toxic materials either absent or in sufficiently low concentrations. A unit operating under equilibrium conditions is referred to as a "balanced digester," one in which the process will continue with a minimum of control. However, if any one of the environmental-control parameters is suddenly altered, or if toxic materials are introduced, the equilibrium is disturbed and an "unbalanced digester" results, that is, one in which the anaerobic process is proceeding at less than normal efficiency. In extreme cases, the efficiency may approach zero, with no gas production; a unit in this condition is known as a "stuck digester."

In this chapter, the major problems encountered in the operation and maintenance of biogas plants will be discussed.

Start-up: Seeding the Biogas Plant

The formation of biogas is a microbiological process achieved primarily by two groups of microorganisms normally present in waste materials such as animal manures.* One group of organisms, the acid formers, is more abundant. The other group, the methane producers, is less abundant, more fastidious, and more susceptible to adverse environmental changes. Waste materials

---

*A third group of microorganisms is responsible for degradation of the waste materials to soluble compounds susceptible to attack by these two groups. See Part II, Chapter 1, Biological Mechanisms.

containing no manure will not have naturally occurring methane producers in large numbers.

In starting up a digester, it is common practice to seed it with an adequate population of both acid-forming and methane-producing bacteria. Generally, actively digesting sludge from a municipal digester, material from a well-rotted manure pit, or cow-dung slurry may be used as the seed to start up a new biogas plant. As a guideline, in continuous or semicontinuous operations the seed material should be at least twice the volume of the fresh manure slurry during the start-up phase, with the daily addition of seed decreased over a 3-week period.

The successful operation of a digester depends on establishing and maintaining a balance between the acid- and methane-forming bacteria. If the digester accumulates volatile acids as a result of overloading, the situation can be corrected by reseeding and temporarily suspending the feeding of the digester, or by the addition of lime. (See Part II, Chapter 1, for a discussion of pH.)

Temperature

Temperature has a significant effect on anaerobic digestion of organic material. The digestion proceeds best at $30°C$-$40°C$ with a mesophilic flora, and at $50°C$-$60°C$ if a thermophilic flora is developed and adapted. The choice of mesophilic or thermophilic operation will normally be made at the design stage and will be based upon climate and other considerations. If considerable energy is required to maintain a digester in the thermophilic range, it may be better to operate at mesophilic temperatures.

Methane-producing bacteria are very sensitive to sudden thermal changes. For optimum process stability, the temperature should be controlled carefully within a narrow range of the selected operating temperature. At the very least, the digester should be protected from sudden temperature changes. It is common practice in many countries to bury the digester in the ground, taking advantage of the insulating properties of the surrounding soil.

To minimize the heating requirements, the insulation of digesters with materials such as leaves, sawdust, and straw, generally available in rural areas, may be warranted. These materials may be composted in an annular ring surrounding the digester, and the heat generated during composting transferred to the digester contents.[1] However, when the temperature decreases as the composting is completed, the composted material should be removed and another batch started. It may also be helpful to bury the digester in a massive compost heap. It should be pointed out, however, that material suitable for composting would prove equally suitable for anaerobic digestion. The trade-off should be evaluated before this heating method is used.

Methane formation is a biochemical process dependent on intimate contact between the microorganisms and the waste material; therefore, it is advantageous to mix the digester contents. Various mixing methods that may be considered are: 1) daily feeding of the digester (semicontinuous operation) instead of filling the digester and removing its contents periodically; 2) manipulating the inlet and outlet arrangements to ensure that the feeding and withdrawal operation encourages mixing; 3) installing mixing devices that can be operated manually or mechanically; 4) creating a flushing action of the slurry through a flush nozzle, as in the German Schmidt-Eggersgluss type of biogas plant;[2] 5) creating a mixing action by flushing the slurry so that the flow is tangential to the digester contents; and 6) installing wooden conical beams that cut into the straw in the scum layer as the surface of the liquid moves up and down during filling and emptying of the digester, as in the German Weber-type design.[2] Where no agitation is provided, stratification in a digester may be minimized by using a horizontal-displacement digester, which simulates a plug type of flow.[3]

Some people working with digesters feel that agitation is desirable but not always essential, since feeding fresh waste continuously produces enough movement in the contents to keep animal manure slurries in suspension. This allows the desired contact between the microorganisms and their substrate. However, mixing may be essential when the organic matter in the digester settles.

In municipal digesters, part of the gas produced is recirculated to mix the digester contents. This practice may be followed in large digesters constructed for biogas production at large-scale livestock operations.

In a pilot-plant study in India, on the digestion of cow dung,[4] daily gas production with gas recirculation was found to be almost twice the amount produced in a conventional digester.

Although there are apparent advantages in recirculating gas to achieve mixing, it is doubtful whether this method can be used in small biogas plants, because the potential operation and maintenance problems may be difficult to resolve in rural areas of developing countries. In larger biogas plants, however, mixing by gas recirculation may be feasible because these plants are more likely to be operated under skilled supervision.

Organic Loading

The quantity of gas produced in a biogas plant per day depends on the amount of waste material fed per unit volume of the digester capacity. The recommended loading rate for standard municipal digesters is from 0.03 to 0.1 lb of volatile solids per ft^3 of digester capacity (0.48–1.6 kg per m^3) per day, and the detention times can vary from 30 to 90 days.

Studies based on dairy cattle wastes in India indicate that at a daily loading rate of 0.42 lb of volatile solids per ft^3 (16.7 kg per m^3), a gas production of 0.62–1.19 ft^3 of gas per lb (0.04–0.074 m^3 per kg) of raw dung fed was obtained.[5]

A recent study reported results on the effect of cow-dung loading on gas production of a pilot-scale dung digester with a capacity of 35.31 ft^3 (1 m^3) and equipped with a 24.5 ft^3 (0.69 m^3) gas holder.[5] These results are given in Table III-4.

From the data in Table III-4 it appears that a loading rate of about 1.5 lb of raw dung per ft^3 per day (24 kg per m^3 per day) is optimal for anaerobic digestion under the conditions of the study. At this loading rate, the maximum quantity of gas per unit weight of volatile solids destroyed was achieved (although the percent reduction of volatile solids was only two-thirds that with smaller loading).

Optimum loading rates for night-soil digesters in various cities of India were reported as ranging from 0.065 to 0.139 lb of volatile solids per ft^3 (1.04 to 2.23 kg per m^3) of digester capacity.[6] The higher loading rates were for towns where the mean ambient temperatures were higher. The comparable volumetric loadings ranged from 1.9 to 0.89 ft^3 per capita per day (0.054 to 0.025 m^3 per capita per day) and were based on a per-capita volatile-solids contribution of 0.123 lb per day (0.056 kg per day). A night-soil loading rate of 0.1 lb volatile solids per ft^3 (1.6 kg per m^3) of digester capacity per day is recommended for temperate climates. An average daily gas production of 0.8–1.2 ft^3 (0.023–0.034 m^3) per capita was recorded for these plants.

Influent Solids Concentration

Gas production from biogas plants is also dependent upon the concentration of solids in the influent slurry. Investigations indicate that, in the absence of toxic materials, optimum gas production is obtained with a 1:1 slurry of cow dung and water.[1,5,7,8] This corresponds to a total solids concentration of about 10–12 percent in the slurry.

Toxicity

A number of substances, both organic and inorganic, may be toxic or inhibitory to the anaerobic process. Tentative guidelines have been developed in the United States for the dilution of various types of manures based on the potential toxicity of ammonia and volatile acids (Table III-5).

TABLE III-4 Effect of Dung Loading on the Gas Production in a Pilot-Scale Dung Digester[a]

Fresh Dung Loaded[b]		Volatile Solids Loaded[b]		Detention Time, days	Volume of Gas Produced[c]		Reduction in Volatile Solids, %	Gas Produced per Unit Weight of Volatile Solids Destroyed	
lb/ft^3/day	kg/m^3/day	lb/ft^3/day	kg/m^3/day		ft^3/day	m^3/day		ft^3/lb	m^3/kg
0.5	8.0	0.073	1.17	50	7.8	0.22	22.8	14.1	0.88
1.0	16.0	0.150	2.40	25	15	0.42	19.5	13.2	0.82
1.5	24.0	0.234	3.76	17	18	0.52	14.5	15.6	0.97
2.0	32.0	0.330	5.29	12	17	0.47	14.2	10.1	0.63

[a]Compiled from the data of Mohanrao.[5] (Percent moisture in dung not given.)
[b]Weight per unit of digester volume per day.
[c]Per m^3 of digester volume.

TABLE III-5 Suggested Dilution Requirements for Animal Wastes[9]

Animal Type	Dilution as Manure/Manure + Water
Swine (growing/finishing)	1 : 2.9[a]
Dairy	Undiluted
Beef, 700 lb (ca. 320 kg)	1 : 2.5[b]
Poultry (layer)	1 : 8.3[a]
Poultry (broiler)	1 : 10.2[b]

[a] Based on ammonia toxicity criteria.
[b] Based on volatile acids toxicity criteria.

In digesters treating poultry wastes, the potential toxicity due to ammonia can be corrected by balancing the C/N ratio of poultry manure by adding shredded straw or bagasse, which have a high C/N ratio. However, it has been reported that the digestion of chicken manure is possible without other supplementary materials. Nevertheless, raw poultry manure must be diluted with water prior to its digestion; otherwise the high concentration of free ammonia produced during initial stages of anaerobic decomposition can be lethal to methane bacteria.

The capacity for a material to inhibit biological activity depends on its concentration. In general, when a substance is present in low concentration, it may be stimulatory, particularly if it is needed in small amounts for biological growth. When its concentration is high, however, it may become toxic. Effective concentrations will of course vary with both the substance and the system. This subject is discussed in detail in Part II, Chapter 1. Materials that can be toxic to anaerobic flora, if present in too high a concentration, include common alkali and alkaline-earth cations, such as sodium, potassium, calcium, and magnesium. Since these are commonly present in chemical fertilizers, care should be taken, where such fertilizers may be in use, to avoid inadvertently introducing them in large amounts.

Some organic materials also inhibit the anaerobic digestion process; these are also discussed in Part II, Chapter 1. Certain precautions should be mentioned here. Residues of crops that have been treated heavily with organic pesticides, such as DDT and other chlorinated hydrocarbons, may introduce toxic inhibitors into the digester. Washing the residues with water before adding them to the digester is not likely to help significantly, since these pesticides are not water-soluble. It may help, however, to mix such crop residues with other feed materials to minimize possible inhibitory effects.

A further source of toxic substances may be the excreta of animals that have been receiving large doses of antibiotics or metals (e.g., copper) as growth factors in their feed or as anthelminthics. These substances may inhibit the growth of bacteria in the digester. As with pesticides, dilution with other, nontoxic, wastes may reduce the concentration of the toxic materials sufficiently to permit the digestion process to continue.

In rural areas of developing countries, however, residues of organic pesti-cides, antibiotics, or anthelminthics would not be expected to be of concern.

Nutrient Deficiency

In anaerobic digestion, a portion of the organic substrate is converted to microbial cells, while most of the remainder is stabilized by conversion to methane and carbon dioxide. The nutritional requirements of the process are related to the quantity of microorganisms produced.

Carbon, hydrogen, nitrogen, and phosphorus are the major elements re-quired for cell growth, although trace amounts of many other elements are also required. Two of the most important nutrients required for anaerobic digestion are organic carbon and nitrogen. To ensure a properly operating process, the carbon:nitrogen ratio of the substrate should never fall outside the range of 30:1–50:1. Where sewage sludge or animal wastes are used as the substrate, this ratio is usually maintained naturally; however, nitrogen defi-ciencies may occur in systems that utilize food-processing wastes or crop residues such as straw. In this case, the carbon:nitrogen ratio may be 100:1 or higher, and supplemental forms of nitrogen, usually in the form of ammonia, should be added.

There is some evidence that the optimum C:N ratio may vary with temper-ature. In winter, for instance, the gas production in cow-dung biogas plants in India dropped to about one-third of the quantity produced during summer months.[10] Gas production could be stimulated in winter, however, by adding easily digestible materials, such as powdered leaves, kitchen wastes, and powdered straw, that promote the multiplication of microorganisms. In this experiment, gas production from 0.5 kg of dung was almost doubled from about 17.2 liters to 31.5 liters at $7°C$ by the addition of 200 ml of urine. It was also reported that dehydrated, urine-soaked materials, such as powdered leaves, straw, and sawdust, retained the property of stimulating gas produc-tion. Thus, where a cattle shed is used, it appears worthwhile to construct a drain for collecting urine and directing it onto materials that can be soaked, dried, and used to feed the digesters. Human urine can be utilized in the same way by collecting it in a pit or a container that can hold soakable materials such as those mentioned above. These materials can be retrieved periodically and dried for use in the biogas plant during the winter months. Use of these materials during summer months also should increase gas production.

Batch or Continuous Operation

In rural settings in a developing country, installation of a truly continuous anaerobic digester is unlikely because of the elaborate feeding and control mechanisms involved. However, a digester may be designed to operate on a "fill and draw" basis with a 1-day to 1-week cycle, or it may be designed on an "all-in, all-out" basis where the reactor is charged and then emptied when gas production is almost completed or at a very low level. A combination of the two systems may also be considered, in which the batch digesters are emptied and charged when a large quantity of waste is generated (for instance, at harvest times). Semicontinuous digesters may be used for waste that is available daily or weekly.

To even out the available energy to meet continuous demands, some gas storage is usually necessary with semicontinuous and batch processes.

Batch processes should be operated in such a way that when the time comes for emptying, approximately one-quarter of the digester contents can be left to seed the incoming batch.

Disposal of Digester Residue

The problems associated with disposal of the spent residue/sludge from anaerobic digestion have been discussed in Part II, Chapters 4 and 5. The major problems involve transportation of the digested slurry. One approach to this problem that avoids this difficult task has been reported from India.[11,12] The spent slurry is carried by a channel from the digester to a sloping filter bed, where it percolates through a 6-in. (15-cm) layer of compacted dry or green leaves. The slope allows the liquid to be partially decanted. The residue can then be handled as a solid or semisolid for transportation to the compost pit. The decanted liquid is then available for mixing with fresh dung to be fed to the digester.

Maintaining the Anaerobic Digestion Process

The balanced digester is one in which the digestion proceeds with a minimum of control. This means that the system environmental control parameters remain naturally within their optimum ranges, with only occasional fluctuations that may require intervention. When an imbalance does occur, the two main problems are the identification of the commencement of the unbalanced condition and the cause of the imbalance. Unfortunately, there is no single parameter that will always indicate the commencement of an unbalanced anaerobic process; several parameters shown in Table III-6 must be monitored simultaneously. None of them can be used individually as a positive indicator of the development of digester imbalance, although decreasing pH and decreasing total gas production certainly are practical indicators.

The most immediate indication of impending operational problems is a significant decrease in the rate of gas production. If the growth of the micro-

TABLE III-6   Indicators of Unbalanced Digestion

Increasing Parameters	Volatile-acid concentration
	Percent of CO_2 in gas
Decreasing Parameters	pH
	Total gas production
	Waste stabilization

organisms is being inhibited by one or more factors, it will be reflected in the total gas production. However, a decrease in the gas production rate may also be caused by a decrease in either the digester temperature or the rate at which the feed material is being added to the digester. Therefore, this parameter can be used as a sure indicator of digester balance only if techniques are available for measuring gas production, maintaining uniform loading of the digester, and monitoring the digester temperature.

The best and most significant single indicator of digester problems would be a change (decrease) in pH. In an operating system, this pH decrease is associated with an increase in volatile-acid concentration. Measurement of the increase in volatile acids is also a good control parameter.

However, laboratory facilities, equipment, and trained personnel are required to monitor most of the control parameters affecting the anaerobic process. This imposes a constraint on the performance evaluation of digesters by farm operators who do not have access to technology or testing facilities.

The biogas-plant operator lacking these facilities must use a more empirical approach to problems of digester imbalance. If a decrease in gas production is noted, the first two things to consider are the temperature and the loading rate. If there has been no sudden change in the ambient temperature—or in the operation of any external heat source used to maintain digester temperature—then it is safe to assume that a temperature change is not the problem. Even without a thermometer, significant temperature changes can be detected. The human hand, in fact, makes a good temperature indicator. It can be used even for digesters operated in the thermophilic range; 140°F (60°C) is just about the upper limit that the hand can withstand without pain or injury. If no significant change in temperature is noted, and the amount and general nature of the feed material has not changed significantly, then a drop in pH, or some toxic material in the digester feed, should be suspected. In that case, the first step is to add lime to treat a pH decrease to see if the gas production begins to rise. If this treatment is not successful, then it is safe to assume that some toxic substance is causing the trouble. In this case, a practical remedy is to dilute the normal feed material with other digestible wastes—straw, hay, grass clippings, urine, night soil—that have not been routinely used during the period of decreased gas production.

Finally, if all else fails and the digester becomes and remains "stuck" (i.e., does not produce methane), the only course of action left is to empty the digester and start again.

Maintenance of the Physical Plant

Animal and human waste is corrosive to metal before, during, and after the anaerobic digestion process. The life of digestion equipment will be extended if tanks and components are fabricated from corrosion-resistant materials. Digester tanks of wood, fiberglass, concrete, masonry brick, or stone will have a longer service life than steel tanks. Metal-tank life can be extended by coating the inside surface with rust-resistant paint, epoxy surfacing, or similar covering. Pipes of stainless steel (or similar corrosion-resistant material) should be used for circulating water through a digester, as in the case of a hot-water heat exchanger to control temperature. Plastics are resistant to corrosion and will give long service when used for low-pressure gas pipelines, low-temperature water lines, and for lines conveying waste materials to and from the digestion units.

Provision should be made to minimize or remove foreign, nonorganic matter from the waste materials to eliminate abrasive action on pumps and mechanical agitators. Appreciable quantities of sand or similar nonbiodegradable materials can lead to troublesome solids accumulation in the bottom of digestion tanks. A screening device should be installed to remove large solids prior to the delivery of waste to the digester units, if this is warranted by the characteristics of the raw waste.

The use of dissimilar metals in water pipes and system components should be avoided to eliminate electrolytic corrosion problems.

It is essential that all components of the digestion system be kept free of gas leaks to eliminate gas loss, accumulation of methane in confined areas (explosion hazard), and the introduction of air (oxygen) to the digester. Routine inspection of all piping and metal components of the digester and gas handling system is necessary to prevent excessive corrosion and maintain the integrity of the process.

References

1. Singh, Ram Bux. 1971. *Bio-gas Plant: Generating Methane from Organic Wastes.* Ajitmal, Etawah (U.P.), India: Gobar Gas Research Station.
2. Joppich, W. 1957. German farms too use fuel gas plants. *Indian Farming* 6(11):35-40.
3. Fry, L. J. 1974. *Practical Building of Methane Power Plants for Rural Energy Independence.* Santa Barbara, California: Standard Printing.

4. Pathak, B. N., Kulkarni, J. M. Dave; and Mohanrao, G. J. 1965. Effect of gas recirculation in a pilot scale cow dung digester. *Environmental Health* (India) VII:208-212.
5. Mohanrao, G. J. 1974. Scientific aspects of cow dung digestion. *Khadi Gramodyog Journal of Rural Economy* (India) 20(April):340-347.
6. _____ n.d. Aspects of Night Soil Digestion, Sewage Farming and Fish Culture. Calcutta, India: All India Institute of Hygiene and Public Health. (Unpublished paper.)
7. Acharya, C. N. 1958. *Preparation of Fuel Gas and Manure by Anaerobic Fermentation of Organic Materials.* Indian Council of Agricultural Research Series, Bulletin No. 15. New Delhi: Indian Council of Agricultural Research.
8. Desai, S. V., and Biswas, S. C. 1945. Manure and gas production by anaerobic fermentation of organic wastes. *Indian Farming* 6(2):67-71.
9. Smith, R. J. 1973. The Anaerobic Digestion of Livestock Wastes and the Prospects for Methane Production. Paper presented at the Midwest Livestock Waste Management Conference, 27-28 November 1973, Iowa State University, Ames, Iowa.
10. Idnani, M. A., and Chawla, O. P. 1969. Improve the performance of your biogas plant in winter. *Indian Farming* 19(6):37-38.
11. Idnani, M. A.; Laura, R. D.; and Chawla, O. P. 1969. Utilizing the spent slurry from cow-dung gas plants. *Indian Farming* 19(1):35.
12. Idnani, M. A., and Chawla, O. P. 1964. Biogas plant slurry is now easier to dispose of. *Indian Farming* 13(11):64.

Chapter 4

Performance Measurement

The performance of a biogas plant is measured by the quantity and quality—i.e., the methane content—of the gas it produces. Detailed information on judging performance has been given in Chapters 1 and 3 of this Part.

Quantity of Gas Produced in Anaerobic Digestion

The quantity of gas produced varies widely, depending upon temperature, loading rate, rate of gas production, detention time, and type of waste material used.* Gas yields are shown in Tables III-1, III-3, and III-4. Reported gas production ranges roughly from 1 to 17 ft^3/lb (0.06 to 1 m^3/kg) of dry volatile solids added. Typical values would be in the range of 6-8 ft^3/lb (0.4-0.5 m^3/kg) of dry volatile solids added, using primarily animal manures, human waste, and crop residue as the feed material, with detention times between 10 and 20 days. The methane content of the gas can be expected to be about 60 percent, which means a methane-production rate of about 3.6-4.8 ft^3/lb (0.22-0.3 m^3/kg) dry volatile solids digested, over an average detention time of about 15 days.**

Quality (Composition) of Gas Produced in Anaerobic Digestion

The composition of the gas produced by a properly functioning anaerobic digester should be about 60-70 percent methane and 30-40 percent carbon

---

*See footnote, Part III, Chapter 1, p. 63.
**Note that the design example in Chapter 2 of this Part was based on a yield of 1 ft^3 of gas produced per pound of fresh dung added to a digester having a detention time of 50 days. At about 20-25 percent dry solids, 1 lb of fresh dung would contain about 0.1-0.15 lb of volatile solids (Tables II-6 and III-4), of which about 0.025-0.038 lb would be destroyed by digestion (Table III-4). This is an effective rate of gas production of 7-10 ft^3/lb dry volatile solids added, or 30-40 ft^3/lb volatile solids destroyed. Since volatile-solids destruction is proportional to detention time, lower gas yields than 1 ft^3/lb fresh dung, at a given temperature, can be expected when the detention time is decreased from 50 days.

dioxide, with a small amount of hydrogen sulfide (approximately 0.1 percent in municipal digester gas) and trace quantities of other gases such as hydrogen, ammonia, and oxides of nitrogen. The composition of the gas is a function of the feed material. A cellulosic waste will produce approximately equal quantities of methane and carbon dioxide. A waste containing protein or fats will produce gas with a higher methane content.

The principal impurities in gas produced from the digestion of most waste materials are carbon dioxide and hydrogen sulfide. The gas as produced has a heat value of 500–700 Btu/ft^3 (18,630–26,080 kJ/m^3) and can be used as a fuel, for heating purposes, or for internal-combustion engines. The contaminants—particularly hydrogen sulfide—cause considerable corrosion in internal-combustion engines. The hydrogen sulfide (H_2S) is oxidized to sulfurous or sulfuric acid, depending on the air-to-fuel ratio. Since both of these products are extremely corrosive to metal parts, removal of H_2S is highly desirable before the gas is used as an engine fuel.

Several methods of CO_2 and H_2S separation can be used, including water scrubbing, caustic scrubbing, solid absorption, liquid absorption, and pressure separation, discussed below.

Water Scrubbing

Water scrubbing may be the simplest method of removing impurities from digester gas. However, water requirements of this process are high. Table III-7 illustrates the solubility of CO_2 in water at various pressures and temperatures. Scrubbing the CO_2 from 7 ft^3 (0.2 m^3) of digester gas at 68°F (20°C) and 1 atmosphere pressure (1.03 kg/cm^2) requires approximately 24.2 gal (91.6 l) of water (assuming gas composition of 35 percent CO_2 and CO_2 density of 0.12341 lb/ft^3 [0.00198 gm/cm^3]).

Increased pressure reduces the water requirement but introduces corrosion problems in the compressor. Further, when appreciable quantities of CO_2 are

TABLE III-7   Approximate Solubility of CO_2 in Water[a]

Pressure		Solubility lb CO_2 per 100 lb H_2O (kg CO_2 per 100 kg H_2O)				
		Temperature °F (°C)				
atm.	kg/cm^2	32(0)	50(10)	68(20)	86(30)	104(40)
1	1.03	0.40	0.25	0.15	0.10	0.10
10	10.3	3.15	2.15	1.30	0.90	0.75
50	51.7	7.70	6.95	6.00	4.80	3.90
100	103	8.00	7.20	6.60	6.00	5.40
200	207	–	7.95	7.20	6.55	6.05

[a]Adapted from Nonhebel (1964).[1]

absorbed, the water generally becomes quite acidic, which poses corrosion problems in handling the spent water. Hydrogen sulfide may also be scrubbed with water but the partial pressure of H_2S in digester gas is normally so low that very little is absorbed in water.

Caustic Scrubbing*

Three agents, NaOH, KOH, and $Ca(OH)_2$, are commonly used in caustic scrubbing of industrial gases that contain CO_2 or H_2S as major impurities. As the solutions are subjected to carbon dioxide in the gas stream, an irreversible carbonate-forming reaction occurs, followed by a reversible bicarbonate-forming reaction

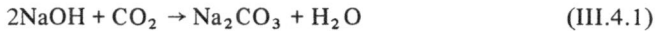

$$2NaOH + CO_2 \rightarrow Na_2CO_3 + H_2O \qquad \text{(III.4.1)}$$
$$Na_2CO_3 + CO_2 + H_2O \rightleftharpoons 2NaHCO_3. \qquad \text{(III.4.2)}$$

In most industrial applications, no attempt is made to regenerate the spent bicarbonate solution because of the high steam requirement for regeneration.

Absorption of carbon dioxide in alkaline solutions is assisted by agitation; promoting turbulence in the liquid aids diffusion of the gas molecules into the body of the liquid and extends the contact time between the liquid and the gas. Another factor governing the rate of absorption is the concentration of the solution. With NaOH for example, the rate is most rapid at normalities of 2.5–3.0.

Although KOH is used extensively in industrial scrubbing, its availability is severely limited in many areas. On the other hand, calcium hydroxide is readily available in most areas, and the cost of operating a lime-water scrubber is minimal. The chief disadvantages in using lime-water are the difficulties in controlling solution strength and removing the large amounts of precipitate $(CaCO_3)$ from the mixing tank and scrubber. It is usually necessary to remove all sediment and suspended particulate matter in order to avoid clogging pumps, high-pressure spray nozzles, and bubbling apparatus or packing.

In areas where sodium hydroxide pellets are available in bulk, their use has the major advantage of enabling rapid and simple recharging of the scrubber, eliminating the problem of suspended particulate matter.

If the contact time is great enough, H_2S can also be removed by caustic scrubbing. It is precipitated by reacting with the carbonate formed in the reaction shown in equation III.4.1:

$$H_2S + Na_2CO_3 \rightarrow NaHS + NaHCO_3. \qquad \text{(III.4.3)}$$

---

*Caustic scrubbing is rarely practiced in small-scale biogas systems. The material in this section is presented for the information of anyone who may encounter excessive CO_2 or H_2S concentrations in large-scale communal or institutional systems.

Hydrogen sulfide and carbon dioxide may also be removed by ammoniacal solutions, trisodium phosphate, sodium phenolate, and other alkacid processes.[2] However, their application to methane recovery appears limited because of the cost of these processes.

Solid-Chemical Absorption

Perhaps the simplest and most economical method of eliminating hydrogen sulfide from digester gas, when other constituents need not be removed, is a dry gas scrubber containing an "iron sponge" made up of ferric oxide mixed with wood shavings.

One bushel (0.0352 m^3) of iron sponge will remove 8.2 lb (3.7 kg) of sulfur.[3] At inlet conditions of 0.2 percent H_2S, this volume of iron sponge will remove the H_2S from approximately 87,000 ft^3 (2,500 m^3) of gas. Regeneration involves exposing the iron sponge to air (i.e., oxygen), which converts the ferric sulfide formed by the scrubbing operation back to ferric oxide and elemental sulfur.

The reactions occurring in this process are:

Scrubbing:

$$Fe_2O_3 + 3H_2S \rightarrow Fe_2S_3 + 3H_2O \qquad\qquad (III.4.4)$$

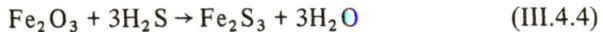

Regeneration:

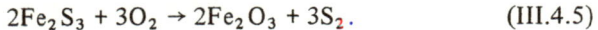

$$2Fe_2S_3 + 3O_2 \rightarrow 2Fe_2O_3 + 3S_2. \qquad\qquad (III.4.5)$$

Minimal expense and maintenance requirements, and ease of regeneration, make the ferric oxide method of H_2S removal a convenient means of protecting storage tanks, compressors, and internal-combustion engines from corrosion caused by prolonged exposure to hydrogen sulfide in the digester gas. Zinc oxide is also effective in removing hydrogen sulfide and has the added advantage of removing organic sulfur compounds such as carbonyl sulfide and mercaptans. Zinc, however, is more expensive than iron, which can be obtained from filings, steel cuttings, and similar sources.

Additional processes are commercially available for separation of the methane from the other gases produced by anaerobic processes. However, these systems are capital- and technology-intensive and are used only when large quantities of gas—1,000 ft^3/min (28 m^3/min) or greater—are available for processing.

References

1. Nonhebel, Gordon, ed. 1964. *Gas Purification Process*. London: George Newnes, Ltd. p. 244.
2. Savery, W. C., and Cruzan, D. C. 1972. Methane recovery from chicken manure. *Journal of the Water Pollution Control Federation* 44:2349-2354.
3. Norris, A. 1943. Scrubbing sewage gas. *Water and Sewage Works* 90:61.

Chapter 5

Safety

Safety concerns related to methane generation involve health hazards and the risk of fire and explosion. The gases predominantly occurring in the environment of a methane-generating system are methane, carbon dioxide, and hydrogen sulfide. Traces of organic compounds such as amines, mercaptans, indoles, and skatoles also occur. Animal wastes may add ammonia to this list. Some properties of the principal noxious gases produced by anaerobic digestion are displayed in Table III-8.

Potentially lethal situations most commonly occur in confined and poorly ventilated operations: a) during manure-pit agitation; b) as a result of ventilation breakdown; and c) during entry into a storage pit. While these situations may not be encountered in a methane-generation plant, they may be confronted in the confinement animal unit for which the methane unit is an auxiliary component. Cautious pit cleanout procedures are the safest way to avoid these problems.

Methane, because of its explosive characteristic when mixed with air, and hydrogen sulfide, because of its potential lethal toxicity, are of major concern in and around a methane-generating system. Hydrogen sulfide content is low (approximately 0.1 percent in municipal digester gas) and has not been found to be a lethal hazard in anaerobic systems used in municipal or farm installations. Methane, on the other hand, is explosive when mixed with air in proportions ranging from 5 to 15 percent by volume. Therefore, rigid precautions against fire and explosion must be exercised in and around a methane-generating plant. Explosion is the major hazard associated with the production, handling, storage, and utilization of methane gas.

A checklist of safety precautions against fire and explosions follows:

1. Prevent digester gas (methane) from discharging into the air in confined areas. This involves gas-tight lines and fittings and pressure-relief valves vented to the exterior of any buildings or confined spaces.

2. Purge air from all delivery lines by allowing gas to flow for an interval prior to use.

3. Install flame traps in gas-delivery lines located in close proximity to the gas-burning appliance.

TABLE III-8 Properties of Noxious Gases and Their Physiological Effects(a)

Gas	Sp. Gr.(b)	Odor	Color	Explosive Range(c)		MIO(d) (ppm)	MAC(e) (ppm)	Concentration(f) (ppm)	Exposure Period(g)	Physiological Effects(h)
				Min. (%)	Max. (%)					
Ammonia (NH$_3$)	0.6	sharp pungent	none	16	–	53	100	400	–	IRRITANT
								700	–	Irritation of throat
								1,700	–	Irritation of eyes
								3,000	30 min.	Coughing and frothing
								5,000	40 min.	Asphyxiating
										Could be fatal
Carbon Dioxide (CO$_2$)	1.5	none	none	–	–	–	5,500	20,000	–	ASPHYXIANT
								30,000	–	Safe
								40,000	30 min.	Increased breathing
								60,000	30 min.	Drowsiness, headaches
								300,000		Heavy, asphyxiating breathing
										Could be fatal
Hydrogen Sulfide (H$_2$S)	1.2	rotten egg smell, nauseating	none	4	46	0.7	20	100	hours	POISON
								200	60 min.	Irritation of eyes & nose
								500	30 min.	Headaches, dizziness
								1,000	–	Nausea, excitement, insomnia
										Unconsciousness, death
Methane (CH$_4$)	0.5	none	none	5	15	–	–	500,000	–	ASPHYXIANT
										Headache, non-toxic

(a) Adapted from Origin, Identification, Concentration and Control of Noxious Gases in Animal Confinement Production Units by E. P. Taiganides and R. K. White, Department of Agricultural Engineering, Ohio State University, Columbus, Ohio. Research Paper of The Ohio State University Research Foundation.
(b) Sp. Gr. = specific gravity: the ratio of the weight of pure gas to standard atmospheric air. If number is less than one the gas is lighter than air; if greater than one it is heavier than air.
(c) Explosive Range: the range within which a mixture of the gas with atmospheric air can explode with a spark (percent is by volume).
(d) MIO = Minimum Identifiable Odor: the threshold odor; i.e., the lowest concentration (highest dilution) from which an odor is detected.
(e) MAC = Maximum Allowable Concentration: the concentration set by health agencies as the maximum in an atmosphere where people work over an 8- to 10-hour period. These levels must be lower in confinement units because animals stay in such environment continuously for 24 hours.
(f) Concentration, in parts of the pure gas in million parts of atmospheric air; to change concentration to percent by volume, divide the listed numbers by 10,000.
(g) Exposure Period: the time during which the effects of the noxious gas are felt by an adult human being and an animal (especially pig) of about 150 lb in weight.
(h) Physiological Effects: those found to occur in adult humans; similar effects would be felt by animals weighing 150 lb; lighter animals will be affected sooner and at lower levels; heavier animals at later times and higher concentrations.

4. Provide adequate ventilation around all gas lines.

5. Provide a vent at the ridge line of the ceiling of buildings to allow the gas, which is lighter than air, to escape.

6. Install gasoline pipes to slope upward or downward with a condensation trap at the low end of the line; digester gas carries water vapor and is described as wet gas.

7. Protect gas lines, particularly condensation and flame traps, from freezing to eliminate the interruption of gas flow, damage to pipelines, and excess pressure buildup in the digester or the low-pressure gas collector.

8. Remove any potential source of sparks or open flames from the gas-production area.

9. Install a fire extinguisher for control of gas fires at the gas-storage location.

10. If pressure storage is used (see Part II, Chapter 3), use storage tanks capable of storing gas safely at pressures up to 2,400 psig (170 kg/cm^2 gauge).

PART IV

RESEARCH
AND
DEVELOPMENT
NEEDS

Chapter 1

General Issues

In the preceding sections of this report, a general overview of the anaerobic digestion process has been given, together with a description of the state of the art and a discussion of the technology involved. Although tens of thousands of biogas plants have been constructed in developing countries, it is not known how many of these are functioning at the expected efficiency. Nor is it certain that the best available materials are being used as feedstocks, or even that they are being used with the best possible preparation and under the most effective conditions.

Several things are clear, however. There is a need in rural areas of developing countries for a cheap fuel that 1) can be used for cooking, heating, and operating irrigation pumps; 2) is under the control of the consumer and need not be purchased; and 3) comes from a renewable local resource. There is also a need for a method of handling human, animal, and agricultural wastes that is economical and attacks the problem of handling and disposal—a problem that is at best a nuisance and at worst a serious public health hazard. Finally, there is a need among farmers in developing countries for an economical fertilizer to maintain or raise the productivity of their land—a fertilizer that can replace the increasingly expensive chemical fertilizers.

The anaerobic digestion process is a promising answer to these needs, but more information is required before this approach can be recommended for large-scale adoption with any assurance of economic success and cultural acceptance.

The section that follows lists the kinds of investigations that could provide the information needed 1) by the planner, in deciding how to allocate scarce construction materials, funds, and technical manpower; and 2) by the technologist, for recommending equipment design and materials-handling procedures.

Survey of the Status of Biogas Plants in Developing Countries

The semiquantitative—and often qualitative—descriptions of biogas generation in developing countries that appear in this and other reports should be
clarified by a detailed survey. The following information would be useful:

- number of digesters constructed
- number of digesters operating
- duration of operation
- number subsidized, and in what way
- comparison of performance of subsidized and nonsubsidized installations
- nature of the digester—individual, community, or institutional
- construction—materials, design, cost
- special problems encountered in construction—site, materials, skills
- operation and maintenance problems, including costs
- who operates the plant
- quality of the gas and residue produced
- amount of energy derived
- use to which the energy is put—cooking, heating, running machinery
- proportion of the energy produced that is actually used
- use to which the residue is put; percent of fertilizer need met
- cultural problems, if any
- evidence of motivation to construct and operate individual biogas
  plants.

Answers to these questions would provide a much firmer basis than do
present reports for preliminary judgment of the potential value of a national
or regional program of biogas production.

Program-Specific Studies

Answers to a different set of questions are also needed before a planning
authority embarks on a national or regional biogas-production scheme.

- What impact will a national or regional policy of building biogas plants
have on the living standards of the population?
  - What is the potential of biogas for meeting the overall energy requirement of the target area?
  - Will the biogas supply energy not heretofore used, thereby making a
direct impact on the standard of living?

— Will it meet a public health need by providing a safe, practical, economical method of handling human wastes?

— Will the fertilizer produced significantly improve agricultural productivity?

• Will it help solve a serious—or potential—deforestation problem by reducing the need for firewood?

• Are there serious sociological or cultural problems to be taken into account?

— Are there taboos that would interfere with the handling of the raw materials or residue or the use of the gas as cooking fuel?

— Are there traditions that affect the choice of individual or community digesters?

• Should individual or community digesters be built?

— Are trained technicians available to operate and maintain community digesters?

— Would community digesters create employment opportunities?

— What would be the consequences of a "stuck" community digester in terms of energy dependence, material flows, and acceptance by the community?

Chapter 2

Technical R&D Issues

Several technical questions must be addressed to improve the prospects of biogas production schemes. They involve research on the nature and properties of the feedstocks, the use of the residue, and new approaches to design and construction of biogas plants.

Improving the Biodegradability of Feedstocks

The degree to which waste materials are susceptible to microbial attack in an anaerobic system should be determined for specific feedstocks available in particular regions. Perhaps even more important is the need for research on pretreatment techniques, commensurate with available resources, that will improve the biodegradability of the locally available materials, and thus improve the efficiency of gas production.

The question of the digestibility of, and gas yield from, mixtures of raw materials—plant, animal, and human wastes—is frequently alluded to in popular works on biogas systems. However, there is not a large body of experience with such mixtures, and coherent research on specific mixtures would contribute greatly to more efficient use of locally available materials. This question also pertains to the various strains of microorganisms that occur in anaerobic systems, and research is needed on development of more efficient strains, or on environmental conditions conducive to greater efficiency of gas production.

Reducing the Costs of Biogas Plants

Experience has shown that the most expensive part of a biogas system is the digester/gas holder. A development program is needed to investigate the use of construction materials less expensive than concrete, sheet steel, and galvanized-iron pipe, and to modify current designs to take advantage of less costly materials. The possibility should be investigated of eliminating the gas holder by techniques of digester construction or programming gas use and system-control parameters.

Using Digester Sludge

The most efficient method of treating and using the effluent from an anaerobic digester depends on the nature of the feedstocks and local conditions. Since a major advantage of biogas plants is the potential fertilizer value of the effluent, an investigation of the handling and use of the effluent should be made under a wide variety of conditions of raw materials, digestion conditions, climate, and soil conditions. This information would be enormously useful in evaluating the contribution of digester effluent to the cost/benefit analysis that must be made before a biogas production scheme is instituted.

APPENDIXES

Construction Materials
for 100 ft³ (3 m³) per Day
Biogas Plant

A Comparison of Two Designs*

The materials required for the construction of two biogas plants, each designed to produce 100 ft³ (3 m³) of gas daily, are presented here, as reported by the designers. Both of these plants have been built in great numbers throughout India. Modifications on the original designs have been made in many cases, but the lists of materials presented here are still useful for comparison.

The design by Acharya[1] was developed at the Indian Agricultural Research Institute (IARI) in 1939. It does not seem to have changed significantly since, and is still being installed not only in India, but also in other countries.[2] The design reported by Singh[3] is an outgrowth of the work begun at the Gobar Gas Research Station in Uttar Pradesh, India, in 1960, based on the earlier work at IARI. The materials for each of these designs are listed below. The designs themselves are illustrated in Part III, Figures III-3 and III-4.

Materials Needed for the Construction of a Biogas Plant that Produces 100 ft³ (3 m³) of Gas per Day (Adapted from Acharya[1] and Singh[2])

Acharya [Refer to Fig. III-3]

1. Excavating equipment, such as shovels, crowbars, etc.
2. 1½″ angle-iron—three pieces 12′ long, bent at 90°, 15″ from one end.
3. Gas holder, 5′ diameter x 4′ high, 16-gauge, mild steel sheet, open at one end (bottom) with three handles (rings) fixed at equal spacing around the closed end and a ½″ diameter rod welded circumferentially around the open end.

---

*The designs are given for illustrative purposes only and no particular recommendation by the panel is implied. See M. A. Sathianathan (1975)[4] for comparison of several types of biogas plant designs.

4. 2½″ galvanized-iron pipe—one piece 12′ long.
5. 1″ galvanized-iron pipe—one piece 4½′ long, one piece 1½′ long.
6. ½″ galvanized-iron pipe—approximately 30′, depending on the distance from the biogas plant to the place where the gas is to be used.
7. ¼″ galvanized-iron pipe—one piece 1′ long (connected to the gas exit at "S").
8. 2½″ galvanized-iron elbow—one piece (connected to Item 4).
9. 1″ galvanized-iron elbow—one piece (for gas pipe "J" joint).
10. ½″ galvanized-iron elbow—four pieces (for gas pipe "J" joints).
11. ½″ galvanized-iron coupling—four pieces (for gas pipe "J" joints).
12. ½″ x ¼″ galvanized-iron reducing coupling—one piece (to be fitted after "S").
13. ½″ x ½″ x 1″ galvanized-iron "tee"—one piece (to be fixed to the gas pipe "J" at "N").
14. ½″ galvanized-iron nipple—two pieces (for gas pipe joints, one at "N").
15. 7″ diameter galvanized-iron pulley—three pieces (to be fixed to the top of angle-irons, Item 2, by means of iron plates, Item 19).
16. ½″ bolts (2″ long, fully threaded) and nuts—12 each (for fixing iron plates, Item 19, on top of angle irons, Item 2).
17. 3/8″ galvanized-iron six-ply twisted wire—45′ (for tying three boxes, Item 18, to the gas holder, over the pulleys, Item 15).
18. Wooden boxes, inner dimensions 15″L x 10″W x 12″H—three (for carrying bricks used as counterpoise weights—at "K").
19. ¼″ iron plate, 1″ x 14″—six pieces [to be twisted midway along the length so that, when mounted in pairs on top of each angle-iron support, they will form parallel surfaces between which a pulley (Item 15) can be mounted; three 5/8″ holes to be drilled in each plate—two for mounting to the angle irons, one for the pulley shaft (Item 20)].
20. ½″ bolts (4″ long) and nuts—three each (to be fixed in Item 19 as pulley shafts).
21. ½″ iron rods, 6′-6″ long—three pieces (to be fixed to the angle-iron supports at "C", as bracing).
22. ½″ nuts—six (to fasten the iron rods, Item 21, to the angle-iron supports).
23. ½″ wheel-cock—one (to be mounted on "J" as the gas exit "S").
24. Copper gauze (screen), 60 mesh—one piece, 2″ diameter (circular) (to be fixed inside the gas outlet pipe "J" before gas exit valve "S").
25. Rubber tubing, 3/8″ inside diameter—5′ (to be connected to the gas exit tube, Item 6, after "S").
26. Burner with rose head—one piece. (The burner is made of a 10-inch piece of 3/8″ inside diameter iron tubing, bent at 90° and fixed between two rectangular blocks of wood. The rose head is made from a 2½-foot length of 7/16″ inside diameter copper tubing, bent into a circular loop with gas-exit holes drilled every half inch.)
27. 16-gauge mild steel sheet, 18″ x 18″—one piece (cover for pit "N").
28. ½″ brass water tap, single-turn—one piece (to be connected to Item 13 via nipple, Item 14).
29. Bamboo—one piece, 7′ long (to be split at one end and forked with a small cross piece, and used for opening and closing the water tap, Item 28).
30. Bricks—2,000 (for constructing the digester, drying beds, etc.).

31. Cement—1 bag (for lining the bottom of the digester "P", for fixing angle-iron supports, Item 2 ("I"), in concrete at "R", etc.).
32. Sand—10 cubic feet.
33. Aggregate (gravel or broken stones)—5 cubic feet (for concrete).

Acharya gives the following description for construction of the gas plant:

To start with, a circular pit of seven feet diameter is dug to a depth of four feet. Then, leaving a ledge R nine inches all round, the pit is continued to be dug with a diameter for 5½ feet only up to a further depth of eight feet (total depth of 12 feet). The flooring and of 4½-inch thickness (sic) are built out of brick and mud (cement or plaster is not necessary in small village installations), till the wall reaches a height of six feet from the bottom. The small platform Q is then built and the cow-dung inlet pipe E is fixed in such a sloping position (after cutting out the earth on one side) that it lies outside the position of the upper four feet wall.

The construction of the wall is then continued right up to the ledge R. At this stage, the angle-irons [are] fitted in cement concrete on the ledge R at equal distances, so that the bent portions are down pointing inwards towards the centre of the tank and the vertical portions are just outside the wall. The gas pipe connection J is also fixed in position on the ledge R, so that the exit pipe rises just outside the wall. The circular wall is then constructed to a height of six inches above ground level, leaving a drain F at one point at the top to carry the digested slurry to the drying bed H, which is 12 feet long and six feet broad, divided into two parallel sections by a central wall along the length.

G is a small pit of inner dimensions 1½ feet x one foot x 1½ feet (depth), constructed at a height of 1½ feet above ground level, at the spot where the upper end of pipe E projects. Cow-dung mixed with an equal quantity of water is poured into the above pit and reaches the bottom of the fermentation tank P.

N is a small pit of inner dimensions nine-inch square and four feet depth, constructed at the point where the "tee" connected to the bottom bend of the gas pipe projects outside the wall. The "tee" carries a water tap which may be opened for a few seconds at intervals of some weeks, with the help of the forked bamboo piece, to let out the moisture accumulated inside the gas pipe J and thus to facilitate passage of the gas. The above pit is covered on top by an iron sheet.

The three wooden boxes K, with holes bored in the sides, are tied to 12 feet lengths of twisted wire, which pass over the pulleys and are tied to the three rings on the top of the gas-drum.

The total floor space required for laying out a plant of the dimensions given here would be about 25 feet x 12 feet, but in case the above floorspace is not available adjoining the house, the drying bed may be omitted and replaced by a slurry pit 3 feet x 3 feet x 3½ feet (depth), in which case the total space required may be cut down to 12 feet x 12 feet.

Singh [Refer to Fig. III-4]

1. Excavating equipment, such as shovels, crowbars, etc.
2. Stone aggregate for making concrete.

3. 3/8" mild steel rods (reinforcement for concrete).
4. 4" galvanized-iron pipe, 17' long—two pieces (for inlet and outlet).
5. 3" galvanized-iron pipe, 8' long—1 piece (gas holder guide).
6. Gas holder, 5' diameter x 4' high, 12-gauge mild steel sheet, open at one end (bottom).
7. 4" galvanized-iron pipe, 8' long (one end to be welded to the inside of the gas holder in the center of the top, the other end placed over guide pipe, Item 5).
8. Cement—40 bags.
9. Sand—300 cubic feet.
10. Brick ballast—100 cubic feet.
11. Bricks—7,500.
12. Insulating material (straw, corn husks, peanut shells, sawdust, etc.).
13. Mild-steel angle iron—100' (for structure and gas-holder guide).
14. ½" alkatheine pipe—50'.
15. ½" alkatheine pipe fittings—bend, elbow, coupling—3 each.
16. 1" alkatheine pipe fittings—bend, elbow, coupling—3 each.
17. Wire gauze 80 mesh—1 square foot.
18. Enamel paint—4 liters.

Singh's description of the single-stage, double-chamber biogas plant to produce 100 cubic feet (3 cubic meters) of gas daily follows:

> This is an underground plant. First a hole should be dug 13' deep and 12' wide. That part of the earth through which the inlet and outlet pipes will pass should be cut away. Around the centre of the hole a concrete base is poured 6" deep, 6' in diameter. The composition of the concrete should [be] 1 part cement, 4 parts sand and 8 parts of 1" stone aggregate, on top of this the digestor will be constructed. It will have a 3" floor and 3" walls, and an inside diameter of 5'-5". Eight M.S. [mild steel] reinforcing rods of 3/8" dia. should be set vertically from base to top, at equal intervals around the digestor wall, and rings should be placed horizontally at 1' distances on top of each other. The cement for the floor and walls will be of composition 1:2:4, using ½" aggregate. The inlet and outlet pipes should be placed and the wall built around it. The concrete must be tightly packed around the pipes to ensure against leakage. The pipes will be 4" G.I. [galvanized iron] pipe and should end about 1-1½' above the digestor floor, and about ¼ of the way inside from the wall. This is so that when the chamber dividing wall is built across the centre of the digestor, the pipes will open right in the middle of each chamber. The wall of the digestor will be built to 4' above ground level. Three feet outside the digestor wall, a brick wall will be built to the same height. In the space between the digestor and this wall, some insulating material will be packed. It would be a good idea to provide some means of descending into this space, in case it should ever be necessary to empty the insulation. Rungs made of M/S rod placed in the digestor wall or extending from the digestor wall to the outer brick wall, will serve this purpose[. B]efore filling the insulation space, the digestor wall should be coated with cement plaster on both sides, i.e., inside and outside, to further seal leaks. Insulation can be provided from materials such as wheat husk or straw, rice husk, corn straw, sawdust, etc. The top of this ring of insulation should be

sealed in some way, to prevent the entrance of water, and the bottom should be left as bare earth or simply covered with bricks to allow for drainage if water should get in any way. Around the part of the brick wall extending above ground, the ground level should be raised to equal the height of the plant, and to provide a walkway of about 3' width. If some insulation coating is available, the digestor floor should be covered with this.

Bisecting the digestor will be a wall 4" thick and 8' high, at the top of which an iron support structure with a guide pipe for the gas holder will be placed. This structure is made of angle iron and the guide pipe is 8' of 3" G.I. pipe. The structure will be set in the digestor wall and solidly fixed on top of the chamber-dividing wall. The pipe must be set at the exact centre of the digestor, because the clearance between the gas holder and the digestor wall is small. In this way, the gas holder will be able to descend into the slurry when empty, and will be even with the raised ground level when full. This requires 4' of vertical travel, thus the top 8' of the plant are left for the gas holder and the bottom 8' of the plant contain the dividing wall.

The mixing tank is a cylinder 2' 4" in diameter and 2' high. Its floor is 1' above the natural ground level to provide enough hydraulic head to feed the plant. The inlet pipe opening must be flush with the bottom of this tank. It can be made of bricks or concrete as is convenient. The volume of this mixing tank is about 8½ cubic ft. which is sufficient for the anticipated daily charge of slurry. The discharge pit should be large enough to accommodate all the spent slurry that is expected to accumulate at any time. It should be lined with bricks or concrete, and the pipe should be just even with ground level. It is this pipe which will determine the level of slurry inside the digestor, as that level cannot rise above the opening of the outlet pipe without the slurry overflowing.

The gas holder is a roofed cylinder of 5' diameter and 4' height constructed of 12 gauge M.S. sheet. It is braced internally with angle irons fitted at different heights, so that when the holder is rotated around the centre support, the surface of the slurry will be agitated and the scum broken up. The central pipe is of 4" G.I. The gas holder will be first riveted, then welded on both sides, and then tested for leaks by filling it with water. After all the leaks are sealed by welding, the holder should be given two coats of enamel paint on both sides. There should be no need to repaint the holder for two years. In the top of the gas holder, there is fitted a small tap and valve of 1" diameter, and to this is connected a flexible pipe coming from the main gas pipe leading to the appliances. Inside the tap from the gas holder, a piece of wire screen is attached to serve as a flame arrester. The top of the gas holder is covered with insulating material. The actual capacity of the drum is less than 100 cubic feet, but if the gas is being regularly used, there is no need to build it much bigger than is called for in this design. For a larger gas plant, the gas holder should accommodate from 50% to 75% of the daily production.

References

1. Acharya, C. N. 1956. Your home needs a gas plant. *Indian Farming* 6(2):27-30.
2. For example: Rahman, Mohibur, and Markon, Pierre. n.d. A Bio-gas Generator at BARD [Bangladesh Academy for Rural Development]. Dacca, Bangladesh: Agri-

cultural Development Agencies in Bangladesh, 549F, Road 14, Dhanmondi, Dacca 5. (Unpublished paper.)

3. Singh, R. B. 1971. *Bio-gas Plant: Generating Methane from Organic Wastes*. Ajitmal, Etawah (U.P.), India· Gobar Gas Research Station.

4. Sathianathan, M. A. 1975. Some designs from abroad. In *Bio-gas Achievements and Challenges*, chapter 9. New Delhi: Association of Voluntary Agencies for Rural Development.

Comparative Costs
and Benefits of
Gobar Gas Plants*

The Khadi and Village Industries Commission in India has instituted a scheme of subsidies and loans to encourage individual families, groups of families, institutions, and communities to construct biogas plants. In addition to capital assistance to individuals, institutions, and cooperative societies, the commission supplies technical guidance for construction, operation, and maintenance. Table A-2-1 lists the details of the financial support. To encourage the integration of latrines in the anaerobic digestion system, the commission offers an additional loan of Rs 400 per latrine. Supplementary loans are also available for gas utilization needs, such as additional piping required, and for use of the gas to generate motive power—the latter at the rate of Rs 1,200 per horsepower of the engine.

An analysis of cost and income for a plant producing 6 m³/day is given in Table A-2-2, based on a comparison for previous use of cattle dung as farmyard manure, or as fuel.

The analysis in Table A-2-2 is based on average dung production of

Buffalo	15 kg per day
Bullock or cow	10 kg per day
Calves	5 kg per day

and gas production of about 1.3 ft³ (0.03 m³) per kilogram of wet dung.

---

*Material drawn from. Fernandez, A., ed. 1976. Gobar gas plant—why and how. *Seva Vani* (May-June):15-21. Further information available from: The Director, Gobar Gas Scheme, Khadi and Village Industries Commission, Irla Road, Vile Parle (W), Bombay 400056, India.

TABLE A-2-1   Capacities and Costs of Biogas Plants

Size of Plant[a]		Estimated Cost as of Feb. 1975 Rupees	Approx. No. Animals Required	Grant (25% of esti- mated cost)	Loan (75% of esti- mated cost)
m^3	ft^3				
2	70	2,332	2-3	583	1,749
3	105	3,016	3-4	754	2,262
4	140	3,360	4-6	840	2,520
6	210	4,175	6-10	1,044	3,132
8	280	5,000	12-15	1,250	3,750
10	350	6,100	16-20	1,525	4,575
15	525	8,500	25-30	2,125	6,375
20	700	11,500	35-40	2,875	8,625
25	875	12,800	40-45	3,400	9,400
35	1,237	18,400	45-55	4,600	13,800
45	1,590	20,740	60-70	5,185	15,555
60	2,120	26,000	85-100	6,500	19,500
85	3,004	38,800	110-140	9,700	29,100
140	4,948	58,000	400-450	14,500	43,500

[a]Volume of gas produced daily.

TABLE A-2-2 Cost/Income Analysis of a 6-m^3/day Biogas Plant[a]

	Basis:	Cost of Gas Holder	Rs 1,670
		Pipeline and Appliances	450
		Civil Construction	2,056
		Total Cost of Plant	Rs 4,176
		Subsidy (25%)	1,044
		Cost Basis	Rs 3,132

	Previous Use of Cattle Dung	
Annual Working Costs	Farmyard Manure	Fuel
Interest on capital @ 15%[b]	Rs 281.88	RS 281.88
Gas holder (10-yr life) 10%	125.25	125.25
Pipeline and appliances (30-yr life) 3.3%	11.14	11.14
Civil work (40-yr life) 2.5%	38.55	38.55
Painting gas holder	100.00	100.00
Maintenance	100.00	100.00
Cost of dung as manure (14.8 tons)	592.00	–
Cost of dung as fuel (in terms of kerosene equivalent @ Rs 1.01/liter)	–	711.00
Total Costs	Rs 1,248.82	Rs 1,367.82
Annual Income		
Manure (22.4 tons @ Rs 50/ton)	1,112.00	1,112.00
Gobar gas (2,190 m^3/yr @ Rs 1.01/ liter kerosene equivalent)	1,371.30	1,371.30
Total Income	Rs 2,483.30	Rs 2,483.30

	Income	Costs	Net Income
Dung formerly used as manure	2,483.30	1,248.82	1,234.48
Dung formerly used as fuel	2,483.30	1,367.82	1,115.48

Notes:

[a]Adapted from Fernandez.

[b]Five equal payments of interest on declining balance.

[c]Note that the loan repayment of Rs 3,132 ÷ 5 = Rs 626.40 will have to be paid annually. For the first 5 years, therefore, the net income for the two bases are Rs 608.08 and Rs 489.08, respectively. However, after repayment of interest installments, the figures increase to Rs 1,516.36 and Rs 1,397.36, respectively.

Comparison of Different Types of Cow Dung Gas Plants[a]

	IARI	
	Unit 1	Unit 2
FARM CHARACTERISTICS		
Size in acres		
(hectares)		
Number of cattle units[b]	10	3
SIZE OF BIOGAS PLANT		
Volume of digester in cubic feet	500	150
(cubic meters)	(14.2)	(4.25)
Volume per cattle unit in cubic feet	50	50
(cubic meters)	(1.4)	(1.4)
GAS PRODUCTION[c]		
Temperature of digester, $^\circ$F	52 (winter)	
($^\circ$C)	(11)	
	85-88 (summer)	
	(29-31)	
Methane content, %	55-60	
Daily production (based on 60% methane) - ft^3	350	100
(m^3)	(10)	(2.8)
Daily production (based on 60% methane)		
per cattle unit - ft^3	35	33
(m^3)	(1)	(1)
Output of gas (60% methane) per unit weight		
of dry matter fermented - ft^3/lb	3.6-8.0[d]	
(m^3/kg)	(0.22-0.5)	
UTILIZATION OF THE GAS - PERCENT		
OF TOTAL CONSUMED		
Household use	100	
Tractor fuel	—	
Production of electricity	—	
Heating digesters	—	

[a] Adapted from Joppich, W. 1957. German farms too use fuel gas plants. *Indian Farming* 6(11):35-40.
[b] One cattle unit = one cow of 1,000-lb weight. A fixed ratio is used for calculating units for other animals in terms of cattle units.
[c] The IARI digesters were unheated; therefore the temperatures are dependent on ambient seasonal conditions. The Schmidt-Eggersgluss and the Weber di-

TYPE OF PLANT

SCHMIDT-EGGERSGLUSS			WEBER	KRONSEDER
Unit 1	Unit 2	Unit 3		
1,360	820	570	50	30
(550)	(330)	(230)	(20)	(12)
280	280	134	26	–
17,600	29,600	14,800	3,530	–
(500)	(840)	(420)	(100)	
63	106	110	136	–
(1.8)	(3)	(3.1)	(3.9)	
82-86	82-86	100-104	86	40-77
(28-30)	(28-30)	(38-40)	(30)	(4.4-25)
	60-63			
10,100	14,600	11,300	1,220	45
(290)	(413)	(320)	(35)	(1.3)
36	52	84	47	
(1)	(1.5)	(2.4)	(1.3)	
3.1-4.7				
(0.2-0.3)				
61	45	–	–	100
20	25	–	–	–
–	–	100	–	–
19	30	–	–	–

gesters were heated; thus, only a single range of temperatures is given for these. The Kronseder digester was unheated and the wide range of temperatures is the result of seasonal ambient fluctuations.

(d)In the original table (Joppich) this was given as 36-8.0, presumably a typographical error.

Sources of Information
on Biogas Plants

Information on biogas plants may be obtained from the following sources:

1. Director Gobar Gas Scheme
    Khadi and Village Industries Commission
    Gramodaya
    Irla Road, Vile Parle (West)
    Bombay—400 056, India

2. Head of the Division of Soil Science and Agricultured Chemistry
    Indian Agricultural Research Institute
    New Dehli—110012, India
        or
    Farm Information Unit
    Directorate of Extension
    Ministry of Agriculture and Irrigation
    New Delhi, India

3. Gobar Gas Research Station
    Ajitmal, Etawah
    Uttar Pradesh, India

4. Director National Environmental Engineering Research Institute
    Nehru Marg
    Nagpur—20, India

5. World Health Organization
    1211 Geneva 27
    Switzerland

6. Economic and Social Commission for Asia and the Pacific (ESCAP)
    Division of Industry, Housing, and Technology
    United Nations Building
    Bangkok 2
    Thailand

7. Bangladesh Academy for Rural Development
    Comilla
    Bangladesh

8. Appropriate Technology Development Organization
    Planning Commission
    Government of Pakistan
    Islamabad
    Pakistan

Glossary

Anaerobic. In the absence of air (i.e., oxygen).

Biogas. The gaseous product obtained by the anaerobic fermentation of organic materials. Since methane is the chief constituent of biogas, the term is often loosely used as synomymous with methane.

Cellulose. Biological polymer composed of sugar molecules; the basic building material of plant fiber.

Cellulolytic. Having the property of hydrolyzing (i.e., making water-soluble) cellulose.

C/N Ratio. The ratio, by weight, of carbon to nitrogen.

Detention Time. The average time that a material remains in the system; generally calculated by dividing the total weight of material in the system by the weight removed per unit time (hour, day, week, etc.).

Digestion. The process by which complex organic molecules are broken down into simpler molecules; in this case the anaerobic process (fermentation) by which bacteria accomplish this decomposition.

Enzyme. Biological catalyst (a protein) that facilitates the breakdown of complex organic molecules into simpler molecules.

Fermentation. The biological process by which organic material is broken down into simpler constituents by microorganisms, usually yeasts; see *Digestion*.

Gobar. Hindi word meaning cow dung.

Lipids. Fatty materials.

Lipolytic. Having the property of hydrolyzing (i.e., making water-soluble) lipid materials.

Methane. The simplest hydrocarbon, consisting of one carbon atom and four hydrogen atoms (CH_4); a flammable, odorless gas.

Methanogenic. Having the property of producing methane.

Night Soil. Human feces.

Phytotoxic. Poisonous to plants.

Proteolytic. Having the property of hydrolyzing (i.e., making water-soluble) proteins.

Rupee. The monetary unit of India, designated by the symbol Rs; $1 equals approximately Rs 8.90.

Substrate. Material supplied for microbial action.

TKN. Total Kjeldahl nitrogen; the amount of nitrogen obtained by the Kjeldahl method of digesting organic material with sulfuric acid.

Total Solids. The weight of the solid matter remaining after a sample is dried
   to constant weight at $103 \pm 1^{\circ}C$.
Volatile Acids. The low molecular weight fatty acids.
Volatile Solids. The portion of solids volatilized at $550 \pm 50^{\circ}C$; the difference
   between the total solids content and the ash remaining after ignition at
   $550 \pm 50^{\circ}C$.

Advisory Committee on Technology Innovation

Members

GEORGE BUGLIARELLO, President, Polytechnic Institute of New York, Brooklyn, New York, *Chairman*

E. R. PARISER, Senior Research Scientist, Department of Nutrition and Food Science, Massachusetts Institute of Technology, Cambridge, Massachusetts

CHARLES A. ROSEN, Staff Scientist, Stanford Research Institute, Menlo Park, California

VIRGINIA WALBOT, Department of Biology, Washington University, St. Louis, Missouri

Board on Science and Technology for International Development

DAVID PIMENTEL, Professor, Department of Entomology and Section of Ecology and Systematics, Cornell University, Ithaca, New York, *Chairman*

GEORGE S. HAMMOND, Foreign Secretary, National Academy of Sciences

RUTH ADAMS, Acting Editor, The Bulletin of The Atomic Scientists, Chicago, Illinois.

EDWARD S. AYENSU, Director, Endangered Species Program, Smithsonian Institution, Washington, D.C.

PEDRO BARBOSA, Department of Entomology, University of Massachusetts, Amherst, Massachusetts

DWIGHT S. BROTHERS, International Economist and Consultant, Fairhaven Hill, Concord, Massachusetts

JOHN H. BRYANT, Director, School of Public Health, Columbia University, New York, New York

GEORGE BUGLIARELLO, President, Polytechnic Institute of New York, Brooklyn, New York

ELIZABETH COLSON, Department of Anthropology, University of California, Berkeley, California

CHARLES DENNISON, Consultant, New York, New York

BREWSTER C. DENNY, Dean, Graduate School of Public Affairs, University of Washington, Seattle, Washington

JAMES P. GRANT, President, Overseas Development Council, Washington, D.C.

WILLIAM A. W. KREBS, Vice President, Arthur D. Little, Inc., Cambridge, Massachusetts

FREDERICK T. MOORE, Economic Advisor, International Bank for Reconstruction and Development, Washington, D.C.

127